Mathematics
for Computer
Programmers

Mathematics for Computer Programmers

CHRISTINE BENEDYK KAY

DeVry Institute of Technology

PRENTICE-HALL, INC., Englewood Cliffs, New Jersey 07632

Library of Congress Cataloging in Publication Data

Kay, Christine Benedyk
  Mathematics for computer programmers.

  Includes index.
    1. Electronic data processing—Mathematics.
2. Electronic digital computers—Programming. I. Title.
QA76.9.M35K38   1984   513   83-19223
ISBN 0-13-562140-2

Editorial/production supervision and interior design: Ellen Denning
Cover design: 20/20 Services, Inc.
Manufacturing buyer: Anthony Caruso

Printed in the United States of America

10   9   8   7   6   5   4   3   2   1

ISBN 0-13-562140-2

PRENTICE-HALL INTERNATIONAL, INC., *London*
PRENTICE-HALL OF AUSTRALIA PTY. LIMITED, *Sydney*
EDITORA PRENTICE-HALL DO BRASIL, LTDA., *Rio de Janeiro*
PRENTICE-HALL CANADA INC., *Toronto*
PRENTICE-HALL OF INDIA PRIVATE LIMITED, *New Delhi*
PRENTICE-HALL OF JAPAN, INC., *Tokyo*
PRENTICE-HALL OF SOUTHEAST ASIA PTE. LTD., *Singapore*
WHITEHALL BOOKS LIMITED, *Wellington, New Zealand*

Dedicated to
Bob
My Husband and Partner

Contents

PART II Algorithms 53

6 FLOWCHARTS

PART III Algebraic Applications for Programming 96

7 LANGUAGE OF ALGEBRA

8 ALGEBRAIC EXPRESSIONS OF "NOT EQUAL"

9 EXPONENTS

PART IV   Advanced Algebra Concepts 137

PART V Binary Systems 246

18   BINARY, OCTAL, AND HEXADECIMAL NUMBERS   278

19   COMPUTER CODES   293

PART VI Boolean Algebra Concepts 304

20   MATHEMATICAL LOGIC   304

21   BOOLEAN ALGEBRA AND COMPUTER LOGIC   322

22   ANSWERS TO ODD-NUMBERED EXERCISES   351

INDEX   387

Preface

The purpose of this book is to present the mathematics needed in programming from the viewpoint of the programmer. As a result, this book focuses on how to interpret a problem and develop a solution algorithm necessary for programming, and not how to do the calculations.

The material is based on lectures for an introductory course in mathematics for computer science students and is organized so it follows the basic steps in developing a program:

The book starts with the basics in learning proramming.
The first steps are developing algorithms and flowcharting.
Sections are organized according to mathematical divisions.
Each section is divided into chapters by specific programming topics.

This material is developed through examples; the problems following each section are similar to the examples. This allows easy cross-reference by the student. In addition, notations used are computer oriented rather than mathematical and thus will aid a student in transferring his/her mathematical skills into programming skills.

I would like to thank my husband, Bob, for his help and support without which this book could not have been written. Thanks also to my Dean, Michael Brown, for giving me the idea for this book; to my colleague, Barbara Harris, for arranging my teaching schedule to allow for writing time; and lastly my thanks to my students at the DeVry Institute of Technology, Chicago, for helping me to test this material.

CHRISTINE B. KAY

Mathematics
for Computer
Programmers

Sets

chapter 1

INTRODUCTION

In all mathematics, everything can be said to belong to a set. The use of sets, as applied to computers, provides a way to describe or identify numbers and/or characters. Therefore, by understanding what a set is, such phrases as "the set of operators" or "the character set of a specific language" will be clear. Sets are used in this chapter so that the general concept can be understood and applied throughout the text.

DEFINITION OF SETS

The term *set* means a collection of objects (numbers, letters, characters, etc.) for which the set definition clearly identifies which objects belong to the set and which do not. When referring to a set, the name of the set is usually expressed as a capital letter, A through Z.

There are two methods of definining specific sets:

1.   By rule definition:

$$A = \text{the set of literals}$$
$$B = \text{the set of positive integers}$$
$$C = \text{the set of numeric operators}$$

or

2.   By list:

$$A = \{a, b, c, d, \ldots, z\}$$
$$B = \{0, 1, 2, 3, 4, \ldots\}$$
$$C = \{+, -, \times, \div, \uparrow \text{ (symbol for raising to a power)}\}$$

Objects belonging to a set are placed within braces.

An object that belongs to a set is said to be a *member* of the set or an *element* of the set. The notation is as follows:

$$a \text{ belongs to set } B$$

therefore,

$$a \text{ is a member of } B$$

or

$$a \text{ is an element of } B$$

or

$$a \in B$$

RELATIONS BETWEEN SETS

In defining sets, there are two overall types of sets that can be referred to: U for the *Universal set* and $\emptyset$ for the *empty set*.

U, the *Universal set*, is one in which all the elements fit a certain classi-fication.

$\emptyset$, the *empty set*, refers to a set that has no elements. For example, the set of all integers between 0 and 1 is an empty set.

A Universal set might refer to the set of positive integers. To identify only a part of the set, such as the set E of even positive integers, the two sets would be

$$U = \{0, 1, 2, 3, 4, 5, 6, \ldots\}$$

and

$$E = \{0, 2, 4, 6, \ldots\}$$

Note that all the elements of set E belong to set U; therefore, E is a *subset* of U. This is written $E \subset U$.

If set W is defined as a set of whole numbers such that $W = \{1, 2, 3, \cdots\}$, then $W \subset U$. In fact, they differ only by one element, 0. A set N can be defined as the set of nonnegative integers such that $N = \{0, 1, 2, 3, \cdots\}$. In this case, $N \subset U$ and in fact N and U have the same number of elements. From these examples, two distinctions can be made regarding subsets: they are either *proper* or *improper*.

Proper subsets are those subsets that have fewer elements than the set in which they are contained.

Improper subsets are those subsets that have the same number of elements as the set in which they are contained. Further, if A is an improper subset of B, then it will be true that $A \subset B$ and $B \subset A$.

Any finite set has a specific number of subsets. To find all these subsets, use the following procedure:

$A = \{\quad\}$ $\qquad\qquad$ $n(A) = 0$

The Subset of A is

$\qquad\{\ \}$ $\qquad\qquad$ (take zero elements at a time)

A has one subset

$B = \{5\}$ $\qquad\qquad$ $n(B) = 1$

Subsets of B are

$\qquad\{5\}$ $\qquad\qquad$ (take one element at a time)

$\qquad\{\ \}$ $\qquad\qquad$ (take zero elements at a time)

B has two subsets

$C = \{0, 1\}$ $n(C) = 2$

Subsets of C are

$\{0, 1\}$ (take two elements at a time)

$\{0\}$ (take one element at a time)

$\{1\}$

$\{\ \}$ (take zero elements at a time)

C has four subsets

$D = \{2, 4, 8\}$ $n(D) = 3$

Subsets of D are

$\{2, 4, 8\}$ (take three elements at a time)

$\{2, 4\}$ (take two elements at a time)

$\{2, 8\}$

$\{4, 8\}$

$\{2\}$ (take one element at a time)

$\{4\}$

$\{8\}$

$\{\ \}$ (take zero elements at a time)

D has eight subsets

Set	A	B	C	D	E	F
n(set)	0	1	2	3	4	5
Number of subsets	1	2	4	8		

The numbers keep increasing by a factor of 2. The pattern implies that for set E, the number of subsets is 8×2, or 16, and for set F, the number of subsets is 16×2, or 32.

EXERCISES

1. List the elements for each set.
   (a) The set of characters in the English alphabet.
   (b) The set of characters in base 2 arithmetic.
   (c) The set of states whose name begins with Q.
   (d) The set of whole numbers that are greater than 2 and less than 12.
   (e) The set of whole numbers that are multiples of 5.
   (f) The set of integers Y such that $Y = 2x + 1$.
   (g) The set of months that have 30 days.
   (h) The set of months that do not have 30 days.

2. Give a rule definition for each set.
   (a) $\{\square, \rule{2cm}{0pt}\}$
   (b) {Sunday, Saturday}
   (c) $\{0, 1, 2\}$
   (d) {a, e, i, o, u}
   (e) $\{1, 3, 5, 7, 9, \ldots\}$
   (f) $\{\ \ \}$

3. List the subsets of {a, e, i, o, u}.

4. Complete the table.

n (set)	1	2	3	4	5	6	7	8	9	10
Number of subsets										

5. Give the formula for finding the number of subsets in a set with n elements.

FINITE AND INFINITE SETS

In certain sets all the elements belonging to the set can be listed, including the final element. For example, in the set A, which is the set of letters in the alphabet,

$$A = \{a, b, c, \ldots, z\}$$

or in the set Q, which is the set of even integers less than 10,

$$Q = \{0, 2, 4, 6, 8\}$$

Both of these sets have a first and a last element; therefore, the number of elements in the sets can be counted. When this is the case, the sets are said

to be *finite*, implying a definite number of elements.   Set Q has five elements, or $n(Q) = 5$. Set A has 26 elements, or $n(A) = 26$.

If a set has no largest or smallest element, it is said to be *infinite*. The
set of positive integers (set P below) is an infinte set because no largest positive integer can be found. The set of negative integers (set P below) has no
smallest element; therefore, the set of negative integers is an infinite set.

$P = \{0, 1, 2, 3, \ldots\}$ The dots (. . .) mean that each
element is one greater than the
preceding element but that the
final element cannot be found.

$N = \{\ldots, -3, -2, -1\}$ The dots (. . .) refer to the remaining
negative integers (progressing to the
left). Again the smallest integer
cannot be found, so the succeeding
elements are formed by adding -1
to each element.

Although two sets may not be equal, they can be *equivalent*.   This
means that the sets have the same number of elements but that the elements
are different.

$A = \{2, 4, 6, 8\}$ $n(A) = 4$; there are four elements in A

$B = \{1, 3, 5, 7\}$ $n(B) = 4$; there are four elements in B

The elements in B are *not the same* as the elements in A, so the sets are
equivalent. If two sets are equivalent, they can be put in one-to-one correspondence with each other.

EXERCISES

1. Identify the sets that are finite and those that are infinite.
   (a) The set of books in the Library of Congress.
   (b) The set of stock numbers identified by a three-digit sequential code.
   (c) The set of numbers divisible by 25.
   (d) The set of days in a week.
   (e) The set of books written by Agatha Christie.
2. Give the number of elements in each set.
   (a) $\{a, b, c, \ldots, z\}$
   (b) $\{10, 11, 12, 13, \ldots, 99\}$
   (c) $\{A0, A1, A2, \ldots, A9, B0, B1, B2, \ldots, Z9\}$
   (d) The set of students in this math class.

(e) The set of states in the United States.

(f) The set of colors in the flag of the United States.

Match statements A to J with sets 3 to 10.

3. _____ {January, June, July}

4. _____ {Y}

5. _____ The set of digits in base 10.

6. _____ { }

7. _____ The set of characters in the word "part."

8. _____ {0, 1}

9. _____ The set of the number of inches in a foot.

10. _____ {10}

    a. The set of consonants that can also be vowels.

    b. The set of numbers Y such that $Y + a = a$ or $Y \times a = a$.

    c. The set of the number of cents in a dime.

    d. The set of integers between 0 and 9, inclusive.

    e. The set of the products of 6×2 and 3×4.

    f. The set of the numbers that are the factors of 10.

    g. The months of the year that begin with J.

    h. The months of the year that begin with K.

    i. The months of the year with 31 days.

    j. The set of characters in the word TRAP.

11. Identify sets that are equal and sets that are equivalent.

    (a) {1, 3, 5, . . .} and {2, 4, 6, . . .}

    (b) {a, e, i, o, u} and {0, 1, 2, 3, 4}

    (c) The set of characters in the word STEAM and the set of characters in the word MEATS.

    (d) The set of characters in the word TRAPPED and the set of characters in the word PARTED.

    (e) {0, 1} and the set of digits in the number 10.

VENN DIAGRAMS

Graphically, a set can be represented by a *Venn diagram*. Although any shape can be used to represent a set, here circles and rectangles will be used for clarity.

Venn diagrams are usually used to show the relationships between the Universal set and its subsets. Because U is the Universal set, it has the largest shape. A is a subset of the Universal set, so it is contained within U and has a smaller shape. If $U = \{1, 2, 3, 4, 5, 6, 7, 8, 9, 10\}$ and $A = \{3, 6, 9\}$, the elements of U and A are added to the diagram, as shown.

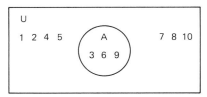

This diagram clearly shows that $A \subset U$ and that $7 \in U$ and $7 \notin A$. Venn diagrams can be used to show set operations.

COMPLEMENTS OF SETS

Complements of a set are those elements that are in the Universal set but do not belong to the subset. x is in the complement of A if $x \in U$ and $x \notin A$. The following Venn diagram is an example.

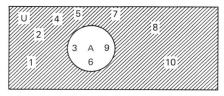

Complements of set A will be those elements that are outside the diagram (circle) of A. Complements of $A = \{1, 2, 4, 5, 7, 8, 10\}$. The notation for a complement of A is $\overline{A}$. If more sets are added to the Universal set, the diagram can be expanded.

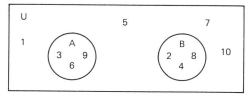

Now the Universal set has two sets, A and B. The complement of A is the set $\{1, 2, 4, 5, 7, 8, 10\}$. The elements in set B are also in the complement of A. $B = \{2, 4, 8\}$.

UNION

The *union* of two sets is the set of elements from both sets. If $x \in A$ or $x \in B$, then $x \in$ of the union of A and B.

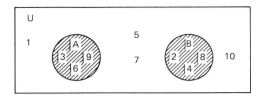

The union of A and B is the set $\{2, 3, 4, 6, 8, 9\}$.

Some sets may share one or more elements:

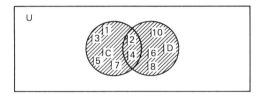

If two sets have common elements, the union of C and D is the set $\{1, 2, 3, 4, 5, 6, 7, 8, 10\}$. The notation for the union of C and D is $C \cup D$.

INTERSECTION

The *intersection* of two sets is the group of elements that are shared by both sets. If $x \in A$ and $x \in B$, then $x \in$ of the intersection of A and B.

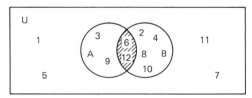

The intersection of A and B is the set $\{6, 12\}$. The notation for A intersect B is $A \cap B$.

The intersection of two sets will contain only the elements common to both sets. However, there is the possibility that the sets will have no common elements. If sets have no common elements, the intersection of the sets is the empty set and the two sets are said to be *disjoint*.

$A \cap B = \emptyset$

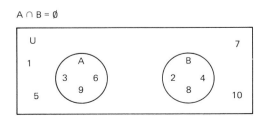

RELATIONSHIPS BETWEEN MULTIPLE SETS

Set operations can be applied to as many sets as can be drawn.

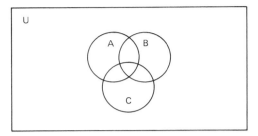

Using Venn diagrams to find $A \cup B$, shade all of circle A and all of circle B.

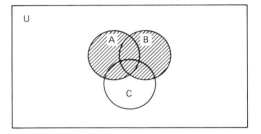

To find the intersection of B and C, shade B with horizontal lines ($\equiv\equiv\equiv$) and shade C with vertical lines ($|\,|\,|\,|\,|\,|$). $B \cap C$ is the area where the vertical and horizontal lines overlap.

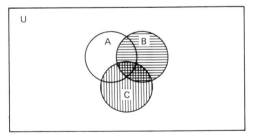

To find $\overline{A}$, shade everything outside A.

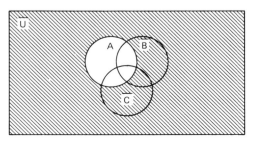

To find $\overline{A \cup B}$, shade everything outside $A \cup B$.

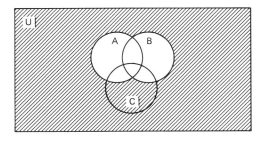

To find $\overline{A \cap B}$, shade everything outside $A \cap B$.

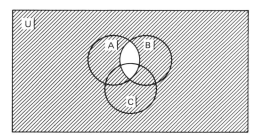

The following represents a combination of operations on sets.

$$(A \cup B) \cap C$$

Shade $A \cup B$ with vertical lines (| | | | |). Since the operation is an intersection, shade C with horizontal lines ($\equiv\!\equiv\!\equiv$). The solution is the area where the horizontal and vertical lines overlap.

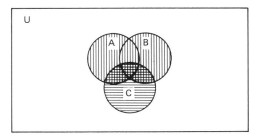

$$(A \cap B) \cap C$$

Shade $A \cap B$ with horizontal lines ($\equiv\!\equiv\!\equiv$). Shade C with vertical lines (| | | | |). The solution is the area where the horizontal and vertical lines overlap.

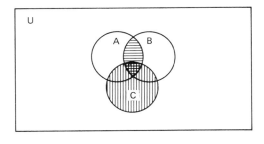

$$(A \cap B) \cup C$$

Shade $A \cap B$ with horizontal lines ($\equiv\!\!\equiv$), and also shade C with horizontal lines. The solution is the total shaded area because the operation is union.

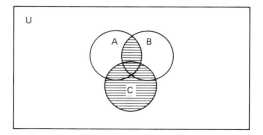

To find the complements, reverse the solutions.

$\overline{(A \cup B) \cap C} =$

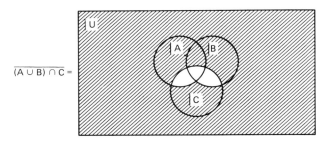

$\overline{(A \cap B) \cap C} =$

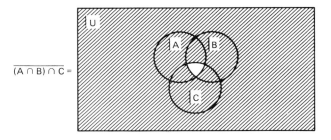

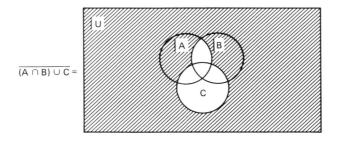

$\overline{(A \cap B) \cup C} =$

EXERCISES

1. Given the sets

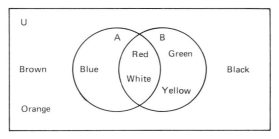

identify each of the following.

(a) U (b) A
(c) B (d) $\overline{A}$
(e) $\overline{B}$ (f) $A \cup B$
(g) $A \cap B$ (h) $\overline{A \cap B}$
(i) $\overline{A \cup B}$

2. Given the sets

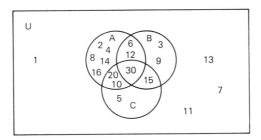

identify each of the following.

(a) U (b) A
(c) B (d) C
(e) $A \cap B$ (f) $B \cap C$
(g) $\overline{C \cap A}$ (h) $A \cup C$
(i) $(A \cap B) \cup C$ (j) $\overline{(A \cup B) \cap C}$
(k) $(A \cap B) \cap C$ (l) $\overline{A \cap B}$
(m) $\overline{A \cup C}$ (n) $(B \cap C) \cup A$

3. Shade each condition on the diagram shown.

(a) $D \cap E$

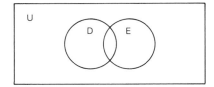

(b) $\overline{B \cup C}$

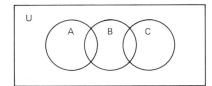

(c) $\overline{(A \cap B) \cup C}$

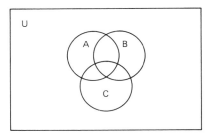

(d) $\overline{(G \cap H) \cap I}$

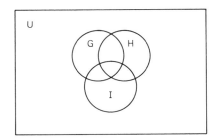

4. Identify each shaded region using set notation.

(a)

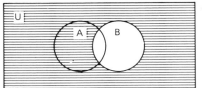

(b)

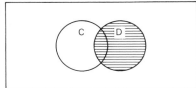

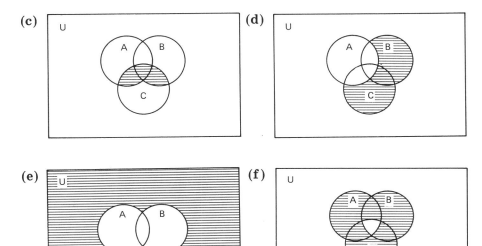

Using Venn diagrams, match the set operations.

5. $(A \cap B) \cup A$ A. $\overline{A \cap B}$

6. $(A \cup B) \cap B$ B. B

7. $A \cup (B \cap C)$ C. $(A \cup B) \cap (A \cup C)$

8. $\overline{A} \cap \overline{B}$ D. $(A \cap B) \cup (A \cap C)$

9. $A \cap (B \cup C)$ E. $\overline{A \cup B}$

10. $\overline{A} \cup \overline{B}$ F. A

SUMMARY

A set is a collection of objects called elements which can be identified by a rule. When this is done, the complement of the set will also be identified as those elements that do not belong to the set.

By comparison, sets may be equal or equivalent to other sets. With equivalent sets, the concept of one-to-one correspondence applies. An example of this concept is setting name cards for seating at a dinner table—for each card there will be a corresponding diner.

Set operations include union, which combines all elements; or intersection, which selects those elements common to more than one set.

A picture of sets and set operations is called a Venn diagram. This kind of diagram allows various set concepts to be illustrated by quickly showing such things as complements, union, or intersection.

C H A P T E R   R E V I E W —————————————————————————————

1. A set is a ————————— of objects.
2. The objects belonging to a set are the ————————— or ——————— ————————— of the set.
3. A set with no elements is the ————————— set, written ——— .
4. Two types of subsets are the ————————— and the ——————— ————————— .
5. If the number of elements in a set and its subsets are the same, the subset is a/an ————————— subset.
6. A set having 16 subsets must have ——— elements.
7. A set that has no largest or smallest elements is said to be ——————— ————————— .
8. If two sets are in one-to-one correspondence, they are ————————— .

Using the diagram

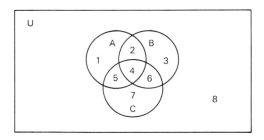

match the items in the two columns.

——— **9.** $\{1, 8\}$ **A.** $\overline{A \cup B \cup C}$
——— **10.** $\{4\}$ **B.** $\overline{A}$
——— **11.** $\{8\}$ **C.** $(A \cap B) \cup C$
——— **12.** $\{4, 5, 6, 7\}$ **D.** $\overline{B \cup C}$
——— **13.** $\{3, 6, 7, 8\}$ **E.** $B \cap (A \cup C)$
——— **14.** $\{2, 4\}$ **F.** $\overline{A \cup C}$
——— **15.** $\{3, 8\}$ **G.** $(A \cap B) \cup (B \cap C)$
——— **16.** $\{1, 5, 7, 8\}$ **H.** $A \cap B \cap C$
——— **17.** $\{2, 4, 5, 6, 7\}$ **I.** $\overline{B}$
——— **18.** $\{7, 8\}$ **J.** $A \cap B$
——— **19.** $\{2, 4, 6\}$ **K.** $\overline{A \cup B}$
——— **20.** $\{2, 4, 6\}$ **L.** C

Integer and Real Number Sets

chapter 2

The two main sets of numbers that are used by programmers are integers and real numbers. All numbers under the four arithmetic operations of addition, subtraction, multiplication, and division can fit into one or the other of these main number sets.

Certain computer languages require that numbers be identified as either integer or real. This chapter deals with the sets of real and integer numbers as required by certain languages, and has as its goal making the student aware of constraints imposed on programmers by other languages.

REAL NUMBERS

The set of *real numbers* can be thought of as a set containing two subsets: rational and irrational. The *rational* set contains the set of integers as its subset, as shown.

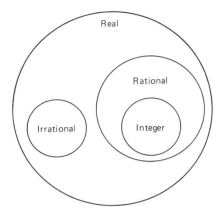

The rational numbers are those numbers that can be represented as a terminating decimal or a nonterminating, repeating decimal. The following are rational numbers:

1.5	terminates at tenths
4.0	terminates at a whole number
1.3333...	nonterminating; the number 3 repeats
1.453453...	nonterminating; the numbers 453 repeat

All *integers* are rational numbers because they are numbers that terminate at a whole number. Integers belong to the real number set and can be treated as real numbers. By adding a decimal point to an integer, an integer becomes a real number. When using real numbers, the numbers will always be treated as decimals. Any fraction can be changed to a decimal, which is the way a computer works with fractions. All fractions are rational numbers because the denominator can divide into the numerator to form a decimal. A fraction such as $\frac{1}{8}$ would become the terminating decimal .125:

$$
\begin{array}{r}
.125 \\
8\overline{\smash{)}1.000} \\
\underline{8} \\
20 \\
\underline{16} \\
40 \\
\underline{40} \\
0
\end{array}
$$

The fraction $\frac{2}{9}$ becomes the nonterminating, repeating decimal .222 . . .:

$$\begin{array}{r} .222\ldots \\ 9\overline{)\,2.000} \\ \underline{1\ 8} \\ 20 \\ \underline{18} \\ 20 \\ \underline{18} \\ 2 \end{array}$$

Irrational numbers are the complements of the set of rational numbers—those numbers that are nonterminating and nonrepeating. One method of finding irrational numbers is to take the square root of numbers that are not perfect squares:

$$\sqrt{2} = 1.4142136\ldots$$

$$\sqrt{3} = 1.7320508\ldots$$

$$\sqrt{5} = 2.236068\ldots$$

The most famous irrational number is $\pi = 3.1415927\ldots$

EXERCISES

1. Identify each number as rational or irrational.
   (a) 7.324324324 ...   (b) 87
   (c) 4.1327890245 ...   (d) $\frac{9}{5}$
   (e) 54.2   (f) $\frac{15}{17}$
   (g) π (h) 5.3333 ...
   (i) $\sqrt{7}$ (j) 411

2. Change each fraction to a decimal number.
   (a) $\frac{5}{11}$ (b) $\frac{8}{5}$
   (c) $\frac{12}{15}$ (d) $\frac{125}{180}$
   (e) $\frac{2}{7}$

CLOSURE AND THE SET OF INTEGERS

The basic arithmetic operations are +, -, ×, and ÷. These operations are performed on the numbers in a specified set of numbers. When this is done, and the result is contained within the set, the set is closed under that operation.

Given the set of integers under addition, multiplication, and subtraction, the result is always an integer. The numbers 2 and 78 are elements of the set of integers:

Under addition: 2 + 78 = 80; 80 is a member of the set.

Under multiplication: 2 × 78 = 156; 156 is a member of the set.

Under subtraction: 2 - 78 = - 76; - 76 is a member of the set.

The set of integers is closed under the operations of addition, subtraction, and multiplication.

Under division there are certain integers divided by integers which yield a result that is not an integer:

$$2 \div 78 = 0.264102641 \ldots$$

The result is a rational number that is nonterminating and repeating. Therefore, under division, the set of integers is not closed.

Thus if numbers are declared integers, then under addition, multiplication, and subtraction, the result will be an integer. As seen in the example above, the same is not true for division. If the numbers are integers, the result can be real.

CLOSURE AND THE SET OF REAL NUMBERS

As we learned earlier, real numbers are composed of rational and irrational numbers. Under the four arithmetic operations of addition, multiplication, subtraction, and division, the results will always be in the set. Therefore, real numbers are closed under these arithmetic operations.

Closure does not exist under all operations. An example of this is the "square root."

$\sqrt{-25}$ is not a real number because the square of any real number— either positive or negative—is positive. For example, 5 × 5 = 25 and - 5 × - 5 = 25.

$\sqrt{-25}$ is an operation in which the solution is not in the set of real numbers. So under the operation of square root, the real numbers are not closed.

Therefore, when a programmer is working with square roots, the operand must be positive in order to find the result.

EXERCISES

1. Classify each number as integer, rational, or irrational.
   (a) 125
   (b) 42.78
   (c) 1.5333 . . .
   (d) 41.252252225 . . .

 (e) 615.0   (f) 72
 (g) 8.415415   (h) 9.28888 . . .
 (i) 11.4687239 . . .   (j) 98

2. State which results are integers.
    (a) 218 + 742   (b) 98 − 176
    (c) 195 ÷ 15   (d) 4 × 23
    (e) 95 ÷ 172   (f) 287 − 99

3. State which results are rational, irrational, or neither.
    (a) 1.243 × 2.7   (b) 8.53 + 17.
    (c) 297. ÷ 3.3   (d) 4.852 − 159.2
    (e) $\sqrt{2.25}$ (f) 78. ÷ 4.
    (g) $\sqrt{17}$ (h) 8.2 + 6.3 ÷ .7
    (i) 198.5 − 7.49 ÷ 6.   (j) 6.13 + $\sqrt{25}$
    (k) $\sqrt{(18. − 78.)}$ (l) 18.4 − $\sqrt{72}$
    (m) 7.543 × 2. + 8.7   (n) $\sqrt{(7.5 ÷ 25.)}$

4. The following are subsets of the real numbers. Under the four operations
    of addition, subtraction, multiplication, and division, test each of the fol-
    lowing for closure, and identify the operations in which the sets are closed.
    (a) 0, 1   (b) 0, 2, 4, 6, 8, 10, . . .
    (c) 1, 3, 5, 7, . . .   (d) −1, −2, −3, −4, . . .
    (e) 0, 3, 6, 9, . . .

INTEGERS: FIXED-POINT NUMBERS

An integer is a real number that has a decimal at the rightmost position. Because the decimal point is always fixed, the integer is called a *fixed-point
number*.

Integers are used in programming in various ways. One way is in specifying integers for a particular program. When this is done, the number of
integer places specified is called the *integer format* (or *number format*).*
Integer format is written as "I*d*," where "I" refers to the integer (a fixed-
point number) and "*d*" refers to the number of spaces the computer should
allow for the digits. For example:

Number	Format	Places
546	I3	5 4 6
1438	I4	1 4 3 8
46	I2	4 6

---

*The integer format, used in a computer program, tells the computer how many
places to allow for the input and output of the number.

If the integer is negative, the format must reserve a space for the negative sign.

Number	Format	Places
– 42	I3	– 4 2
– 1732	I5	– 1 7 3 2
– 9	I2	– 9

If the integer format is larger than the number given, the digits are right-justified (starting at the far-right position and working toward the left).

Number	Format	Places
13	I5	_ _ _ 1 3
4	I4	_ _ _ 4
– 75	I6	_ _ _ – 7 5

When the integer format is too small for the number, the leftmost digits are lost.

Number	Format	Places	
1243	I3	2 4 3	error
57293	I4	7 2 9 3	error
– 892	I3	8 9 2	error

EXERCISES

1. Specify the integer format for each number.
   (a) 78   (b) 4872
   (c) – 935   (d) 87432
   (e) – 5   (f) – 11428
   (g) 185   (h) – 884325
   (i) 566437   (j) – 566437

2. Fill in the number according to each format. Use the integer – 12345678. Specify "error" where the format yields an incorrect result.
   (a) I10   _ _ _ _ _ _ _ _ _ _
   (b) I9   _ _ _ _ _ _ _ _ _
   (c) I5   _ _ _ _ _

    (d) I8   _ _ _ _ _ _ _ _

    (e) I12   _ _ _ _ _ _ _ _ _ _ _ _

3. Fill in the numbers for each format. Specify "error" where the format yields an incorrect result.

	Number	Format	Places
(a)	4325	I4	_ _ _ _
(b)	- 238	I4	_ _ _ _
(c)	- 5432	I5	_ _ _ _ _
(d)	8900000	I8	_ _ _ _ _ _ _
(e)	- 5400	I6	_ _ _ _ _ _
(f)	- 114930000	I10	_ _ _ _ _ _ _ _ _ _
(g)	- 7134	I8	_ _ _ _ _ _ _
(h)	4000000	I5	_ _ _ _ _
(i)	1040603	I6	_ _ _ _ _ _
(j)	- 4880335	I6	

4. Give the integer format for each quantity.

    (a) 10 gross   (b) Zip code

    (c) Inches in 5 yards   (d) Area code

    (e) Days in 4 years   (f) Social Security number

    (g) The current year   (h) Hours in a week

    (i) Telephone number   (j) Days in 5 weeks

REAL NUMBERS: FLOATING-POINT NUMBERS

A real number is a decimal number, where the decimal point can be in any position in the number. These numbers are *floating-point numbers.*

    The format for real numbers is "Fw.d," where "F" is for floating-point number; "w" is for the total number of places for the number, *including the sign and the decimal point*; and "d" is for the number of decimal places to the right of the decimal point. It will always be true that "w" is greater than "d". The format is illustrated as follows:

If   Fw.d = F5.2

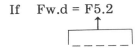

5 represents the total number of places, including the sign and decimal point

If   Fw.d = F5.2

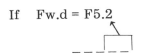

two places are reserved for the decimal positions.

The following table illustrates further how the format works:

Number	Format	Places
4.3	F3.1	4 . 3
- 7.453	F6.3	- 7 . 4 5 3
87.	F3.0	8 7 .
.8582	F5.4	. 8 5 8 2
.98	F1.2	—

error—width of field is too small for number of decimal places

The last number in the table shows that an error will occur when the format is specified incorrectly. Here is another example of format error:

$$73.456 \qquad F5.3 \qquad 3 . 4 5 6$$

The digit 7 was dropped because the number of places did not include a place for the decimal point.

Here is another type of format error:

$$- 5.003 \qquad F6.2 \qquad \_ - 5 . 0 0$$

In this case, the number of decimal places were too few for the number. As a result, the "3" in the decimal portion of the number was lost.

Incorrectly setting the width and the decimal positions will result in one of three consequences:

1. If the decimal positions in the format are less than the decimal position in the number, the decimal digits in the result will be *rounded* to match the format. (See the first six lines in the following table.)

2. If the decimal positions in the format are greater than the decimal positions in the number, significant digits (digits to the left of the decimal point) are lost and an error occurs. (See the last line in the table.)

Number	Format	Places
42.726408	F8.0	_ _ _ _ _ 4 3 .
	F8.1	_ _ _ _ 4 2 . 7
	F8.2	_ _ _ 4 2 . 7 3
	F8.3	_ _ 4 2 . 7 2 6
	F8.4	_ 4 2 . 7 2 6 4
	F8.5	4 2 . 7 2 6 4 1
	F8.7	. 7 2 6 4 0 8 0

error—significant digits lost

3. If the width of the number in the format is too small but the decimal positions are correct, again significant digits will be lost:

42.726408   F8.6 $\underline{2} \underline{.} \underline{7} \underline{2} \underline{6} \underline{4} \underline{0} \underline{8}$ *error*

When number quantities are specified in general terms, the floating-point format should fit the largest possible number:

1. *Quantity: The cost of a week's groceries.* The largest amount spent would probably be in the hundreds, so the largest amount would be $999.99. The format will be of a width greater than or equal to 6, and with a decimal position of 2. The format is F6.2.

2. *Quantity: The number of years in 18 months.* Change months to years by dividing 18 by 12. $18 \div 12 = 1.5$. The format is F3.1.

INTEGER-TO-REAL CONVERSION

Using formats, numbers can be given as integers and changed to real numbers. Since an integer is a number without a decimal point, changing it to a real number means adding a decimal point to the number.

Number	Format	Places
4832	F5.0	$\underline{4} \underline{8} \underline{3} \underline{2} \underline{.}$
	F5.1	$\underline{4} \underline{8} \underline{3} \underline{.} \underline{2}$
	F5.4	$\underline{.} \underline{4} \underline{8} \underline{3} \underline{2}$

EXERCISES

1. Write the format for each floating-point number.
   (a) 68.2   (b) – 7.124
   (c) .00052   (d) 999.01
   (e) – 823.476023   (f) 154.8402
   (g) – .00000019   (h) 9000000.

2. Write each number according to the format specified.
   (a) 6.823   F5.3   (b) 95.0003   F7.4
   (c) – 78.0333   F8.4   (d) 12687.   F6.0
   (e) – 47452.811   F10.2   (f) – 592.   F9.2
   (g) 11.7683333   F10.5   (h) .000042   F8.7
   (i) 17000000.002   F12.4

3. Write the number 858.00073942 according to each format.
   (a) F12.5   (b) F12.7
   (c) F12.8   (d) F12.9
   (e) F12.10   (f) F12.3
   (g) F10.8   (h) F10.7
   (i) F10.6   (j) F10.5
   (k) F10.4

4. Give the floating-point-number format for each quantity.
   (a) $\frac{4}{25}$ (b) $12\frac{1}{2}$
   (c) Five quarters   (d) The cost of this book
   (e) The cost of a new car   (f) The cost of the Sunday paper
   (g) $\sqrt{1600}$.   (h) Grade-point average

5. Given the integer 948034, change it to a real number according to each format.
   (a) F7.0   (b) F8.0
   (c) F8.2   (d) F7.6
   (e) F8.5

REAL NUMBERS: EXPONENTIAL FORM

Real numbers that are very large may be too large to use in the computer. They are very awkward to work with, but can be made easier to deal with by putting them in exponential form. The exponential form of a number is

$$.n \times 10^e$$

where $.n$ is the number and e is the power of 10 that denotes its decimal position.

A number such as 500000000. can be changed to exponential form by shifting the decimal either right or left so that it will be before the first significant digit. The places are then rewritten as a power of 10. For 500000000. the first significant digit is 5 and the decimal point was shifted nine places to the left. The number in exponential form is $.5 \times 10^9$.

For a number such as .007888, the first significant digit is 7. The decimal is shifted two places to the right and the number becomes: $.7888 \times 10^{-2}$.

As indicated above, shifting to the left will give a positive power of 10, and shifting to the right will give a negative power of 10. To illustrate further:

Number	Exponential form
6800000.	$.68 \times 10^7$
.0003	$.3 \times 10^{-3}$
$-52000.$	$-.52 \times 10^5$
$-.028$	$-.28 \times 10^{-1}$

Zeros are included in the number if the zero is between significant digits; otherwise, the zeros are dropped. For example:

Number	Exponential form
.00905	$.905 \times 10^{-2}$
89500000	$.895 \times 10^{8}$

In the first line, the zero between the significant digits 9 and 5 is included. In the second line, the zeros are not between significant digits; therefore, they are dropped.

In most computer languages the exponential form uses an E to replace the "$\times 10$": for example,

$$.905 \times 10^{-2} \quad \text{is} \quad .905E-2$$

$$.895 \times 10^{8} \quad \text{is} \quad .895E+8$$

When using the exponential form of a number, the format is "Ew.d," where E identifies the exponential format; w identifies the width of number places, including sign, E, and decimal point; and d identifies the number of decimal places.

Ew.d = E5.1

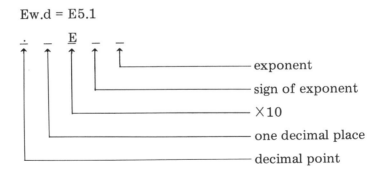

Here are some examples of Ew.d:

Number	Format	Places
.905E-2	E7.3	$\underset{.}{} \underline{9} \, \underline{0} \, \underline{5} \, \underline{E} \, \underline{-} \, \underline{2}$
.8E+11	E6.1	$\underset{.}{} \underline{8} \, \underline{E} \, \underline{+} \, \underline{1} \, \underline{1}$

If the numbers are not given in exponential form to determine their format, they must first be converted to exponential form:

Number	Exponential form	Format
70350000	.7035E+8	E8.4
400000000000	.4E+12	E6.1
$.25 \times 10^{-4}$	.25E-4	E6.2

EXERCISES

1. Rewrite in exponential form using $.n \times 10^e$.
   (a) .000000782   (b) 1200000000
   (c) 7.943   (d) .000005003
   (e) 60040000
2. Rewrite without exponents.
   (a) $.44 \times 10^5$ (b) $.302 \times 10^{-4}$
   (c) $.75003 \times 10^8$ (d) $.611 \times 10^{12}$
   (e) $.184 \times 10^{-9}$
3. Give the exponential formats.
   (a) .982E+5   (b) .76E-9
   (c) $.1004 \times 10^{-4}$ (d) $.54 \times 10^8$
   (e) $.2 \times 10^{12}$ (f) 890000000
   (g) .000002001   (h) .00045
   (i) 23000000   (j) 640000000

SUMMARY

As shown in this chapter, the programmer can work with fractions and integers by changing them to real numbers, and with numbers too large (or too small) for the computer by using exponential notation. The only constraint in working with real numbers is that the square root of negatives cannot be taken. Therefore, attempting to take the square root of a negative will result in an error message.

In most computer languages (and many hand-held calculators), real numbers are referred to as floating-point numbers. Integers are referred to as fixed-point numbers.

Formats are important because some program languages require number specifications for input and output. Formats that do not match the input or output number will result in output error or incorrect results.

To summarize the information given in this chapter dealing with the real number system, the following chart is presented:

Reals	Terminating, nonrepeating	Nonterminating, repeating	Nonterminating, nonrepeating	Fixed-point	Floating-point
Irrational			X		X
Rational	X	X			X
Integers	X			X	

CHAPTER REVIEW

1. The set of real numbers is made up of two subsets: _____ and _____ .

2. The set of _____ is a subset of the set of rational numbers.

3. A nonterminating and nonrepeating decimal is said to be a(n) _____ _____ .

4. A terminating decimal is said to be a(n) _____ .

5. A nonterminating and repeating decimal is said to be a(n) _____ _____ .

6. A fraction is changed into a decimal by dividing _____ by _____ .

7. The square root of a number that is not a perfect square is a(n) _____ number.

8. The most famous irrational number is _____ .

9. If, under an operation on the elements of a set, the result is in the set, then the set is _____ for that operation.

10. The set of integers is not closed for the operation _____ .

11. The set of real numbers is closed under _____ , _____ , _____ , and _____ .

12. The set of real numbers is not closed for a square root because the square root of a _____ number is not in the set of real numbers.

13. Integers are also known as _____-point numbers, and real numbers are known as _____-point numbers.

14. A format of I5 means a ——————————— number having —————— digits.

15. A format of F7.3 means a(n) ————-place number having —————— decimal places.

16. To change an integer to a real number involves adding a ——————————— ——————————— to the number.

17. In a floating-point format, the width is always ——————————— than the number of digits in the number.

18. Very large numbers are changed to ——————————— form.

19. The E in .7E-3 means ————— . The -3 is the ——————————— .

20. In the format Ew.d, w must include a space for a ——————————— , ——————————— , and the ——————————— of the exponent.

Format Arithmetic

chapter 3 _____

The computer is used to do the tedious work of arithmetic. But to direct the computer properly in this function, the programmer must understand the number fields of an operation, and the size of the result. The reason for this is that in many computer languages, the programmer must specify the size of the result fields.

It is not always possible to know the exact size of a result beforehand, but it is possible to make an "educated guess," which is necessary from a programming standpoint. This chapter presents a way to make this educated guess.

ADDITION OF FIXED-POINT NUMBERS

In arithmetic operations it is possible to specify the result field by simply knowing the sizes of the operands. In addition, if the size of the addends are given in fixed-point format, a field can be established for the largest possible result by allowing for a *carry*.

In adding two numbers having the same format, the result can at most be one more than an addend format. The extra position* is for the carry

_____
*The words "position" and "place" mean the same thing and are used interchangeably in this text.

that can occur. For example: Add two numbers having a format of I4 each. Use the format I5 for the result.

         9999   (the largest integer with four digits)
       + 9999
        19998
        ↑_____ carry   The result is five digits.
                                I5 is a sufficient format.

When adding two numbers having variable formats, the result can be at most one more than the largest format. For example: Add two numbers, one having a format of I6 and the other having a format of I8. For the result, select the format I9.

        99999999   (largest possible integer having eight digits)
      +   999999   (largest possible integer having six digits)
       100999998   (nine-digit result)

Negative integers operate in the same way. Add two negative integers, each with a format of I3. The result field should be I4. For example:

       - 99   I3
      + - 99   I3
       - 198   I4   The result is one digit more than an addend.

With two integers that have the same format, but one is negative and one is positive, the result would follow the same rule:

       - 99   I3
      + 999   I3
      _9 0 0   I4   (no carry occurs, so the result has a blank)

In adding a list of numbers with a fixed format, the following procedure should be used. If there are 10 numbers in the list, each having a format of I3, then a result field of I4 would be large enough.

         999   I3
       ✕ 10   (there are 10 numbers in the list)
         9990   (the result will be, at most, I4)

If the list has 100 numbers, the result will need two more places, or I5.

         999   I3
       ✕ 100   (100 numbers in the list)
        99900   (largest number possible is I5)

With variable-length numbers, apply the same rules to the addend with the largest format.

SUBTRACTION OF FIXED-POINT NUMBERS

The format for dealing with subtraction of fixed-point numbers is

$$\text{operator } 1 - \text{operator } 2 = \text{result}$$

If operator 1 is greater than operator 2, use a result format having *one more place* than operator 1. This will accommodate those cases where operator 2 is negative. The result for operators having a format of I3 is I4. For example:

$$999 - 100 = \underline{\ 8\ 9\ 9}$$

$$999 - (-99) = \underline{1\ 0\ 9\ 8} \qquad \text{(the extra position is needed for the carry, because the second operator is negative, which changes this problem to addition)}$$

If operator 2 is greater than operator 1, the extra position will be needed for a negative sign.

$$\underline{1\ 0\ 0} - \underline{9\ 9\ 9} = \underline{-\ 8\ 9\ 9}$$

MULTIPLICATION OF FIXED-POINT NUMBERS

The result field in multiplication of fixed-point numbers is the *sum of the two operator fields*. An integer of I3, multiplied by an integer of I4, results in a product of at most 17.

9999	(largest integers of four digits)
$\times\ \ 999$	(largest integers of three digits)
9989001	(result is seven digits; format I7)

EXERCISES

1. Give the result field format.
    (a) I5 + I5 (b) I4 + I7
    (c) I9 – I4 (d) I8 – I7
    (e) I4 × I2 (f) I9 × I3
    (g) (I4 + I5) + I3 (h) (I2 × I2) + I4
    (i) (I9 – I7) × I2

2. Give the result field format.
   (a) 9483 + 820   (b) 7721 − 9
   (c) 1192 + 98 + 1012 + 4   (d) 187 × 23
   (e) 1235 × 110   (f) 9873 × 9

3. For each of the following, show (1) the format for the given information, and (2) the format of the result. Work out the problems and verify your results.
   (a) Calculate the sum of these class averages: 98, 72, 88, 65, 93, 84, 88, 70, 76.
   (b) Calculate the number of cans in 15 dozen.
   (c) Calculate the number of sale items left if the original amount in stock was 100 gross (1 gross equals 12 dozen), and the number of items sold in 3 days was 3560.
   (d) In a certain class, there were 20 homework assignments each worth 10 points, 10 quizzes each worth 15 points, and three tests each worth 50 points. What is the maximum number of points a student can receive?
   (e) A theater sold 1252 seats for a performance. The theater's capacity is 1500 seats. How many seats were unoccupied?

FLOATING-POINT ADDITION AND SUBTRACTION

When adding two decimals, the basic rule is to line up the decimal points, one under the other. In adding numbers with formats of F5.2 and F4.1, the sum must include decimal places and a possible carry in the leftmost digit position. For example:

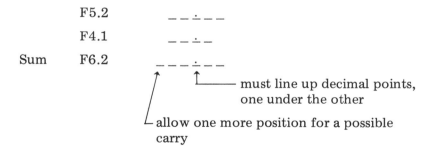

In this operation, both parts of the number must be considered: the whole number and its decimal positions. The number of decimal positions in the result (sum) *must match* the largest number of decimal positions of any of the addends. The number of whole-number positions in the result *must be one more than the largest number of whole-number positions* of any of the addends. For example, to add numbers with F7.4 and F6.3, the positioning would be as follows:

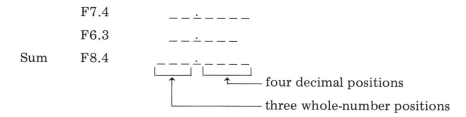

       F7.4

       F6.3

Sum   F8.4

four decimal positions

three whole-number positions

This yields a sum of F8.4.

    In subtraction, it is also important to line up the decimal points, one under the other just as in addition. Therefore, in the subtraction operation of F5.3 – F4.2, the result is

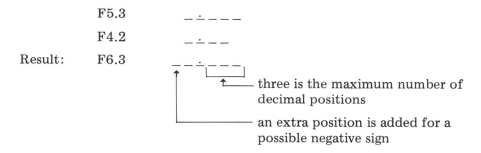

       F5.3

       F4.2

Result: F6.3

three is the maximum number of decimal positions

an extra position is added for a possible negative sign

The rule for subtraction is similar to the one used for addition.

FLOATING-POINT MULTIPLICATION AND DIVISION

The multiplication of decimals will yield a result that must take into account the number of whole-number places and the number of decimal places. With this in mind when multiplying two decimals, the number of decimal places in both multipliers are added. The total number of places in the result will be the sum of the decimal places, plus the sum of the whole-number places, plus one additional place for the decimal point. To illustrate:

$$F9.5 \times F7.3 = F15.8$$

The sum of the decimal places is 3 + 5 = 8.

The sum of the whole-number places is 3 + 3 = 6. (F9.5: one decimal-point place, plus five decimal places, equals six places. Six places subtracted from a total of nine places leaves three places.)

The total number is 8 + 6, which equals 14 places.

Adding one place for the decimal point makes the total 15 places.

The result is F15.8.

Proving the Rule

When multiplying an F5.3 number by an F3.1 number, the result can be, at most, F7.4.

	F5.3	9 . 9 9 9
	F3.1	✕ 9 . 9
Result:	F7.4	9 8 . 9 9 0 1

The F7.4 result was calculated by using the procedure outlined above. In this example the number 9 was used in the multipliers purposely because it is the largest single-digit number. In this way, the result can be expected to use the maximum number of places. Yet the numerical answer has the same number of places as indicated by the F7.4 format.

The division of two floating-point numbers may not yield an exact answer. When this occurs, it is because there is a remainder (R). For example:

$$\frac{17890.}{5.\overline{)89453.}} \quad (R = 3)$$

In analyzing this operation, two conclusions are made: (1) the result field can be at most as large as the dividend field, and (2) the remainder will always be a smaller number than the divisor. The example given earlier, expressed in a floating-point-number format, would be F6.0 ÷ F2.0. Therefore,

The result field is F6.0.

The remainder for a one-digit divisor (F2.0) can be, at most, a one-digit number.

Another example: F5.3 ÷ F2.0. A sample problem would be

$$\frac{2.371}{4.\overline{)9.486}} \quad (R = 2)$$

The quotient is 2.371. The remainder is the digit 2. Because the dividend (9.486) has decimal positions, *the remainder will have an equal number of decimal positions*. To find its decimal position, use this formula:

$$(\text{result} \times \text{divisor}) + \text{remainder} = \text{dividend}$$

In this problem:

Result: 2.371

$\times$ divisor: $\times \underline{\qquad 4.}$

                            9.484

+ remainder: $+ \underline{\quad .002}$ the remainder 2 is added to the
                          9.486   rightmost digit place

For this example of F5.3 ÷ F2.0, the result field is F5.3 and the remainder field is F4.3.

If the divisor is a number with decimal places, certain rules that apply to division by decimals must be followed.

```
            11.1
      8.5 | 94.83      The decimal point in the divisor must be
           85          shifted one place to the right to change
           98          the divisor to a whole number.  The same
           85          number of places must be shifted to the
          133          right in the dividend.
           85
remainder:  48
```

The original floating-point-number format was F5.2 ÷ F3.1. After shifting the decimal points, the format becomes F5.1 ÷ F3.0. The result field is F5.1 (The result field can be, at most, as large as the dividend field.) Following the formula given earlier, we have

Result   11.1
$\times$ divisor $\times \underline{\;\; 8.5}$
                       555
                 $\underline{888\;}$
                94.35
+ remainder: $\underline{\;\;.48}$
                94.83

The remainder field must have two decimal places (equal to the dividend before shifting) and can have at most two digits. The field size would be F3.2.

To divide two numbers where the number of decimal positions in the divisor is greater than the number of decimal positions in the dividend, follow this procedure:

To divide 863.44 by .111 (or to divide F6.2 by F4.3), shift the decimal point three places to the right, and add another position (0) in the dividend.

$$\underline{8}\ \underline{6}\ \underline{3}\ \underline{4}\ \underline{4}\ \underline{0}\ \underline{.}\ \div\ \underline{1}\ \underline{1}\ \underline{1}\ \underline{.}$$

the result field is F7.0

The remainder field can have at most three digits. Since the number of decimal places in the divisor is greater than the number of decimal places in the dividend, use the decimal places in the divisor.

$$\underline{.}\ \underline{-}\ \underline{-}\ \underline{-}$$

three digits—four decimal places

The remainder field is F4.3.

For this problem:

```
                  7778.
        .111 │ 863.440
               777
               864
               777
               874
               777
               970
               888
                82
```

Following the formula given earlier, we have

```
Result:          7778
× divisor:      × .111
                 7778
                 7778
                 7778
               863.355
+ remainder:      .082
               863.440
```

To divide 654.2 ÷ 43.121 (F5.1 ÷ F6.3):

1. Shift three places.

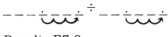

Result: F7.0

2. The remainder at most will have five digits and three decimal places. The format is F6.3.

To verify:

$$
\begin{array}{r}
15. \\
43.121\overline{)654.200} \\
\underline{43121} \\
222990 \\
\underline{215605} \\
7385
\end{array}
$$

$$
\begin{array}{rr}
\text{Result:} & 15. \\
\times \text{ divisor:} & \times\,43.121 \\
\hline
& 15 \\
& 30 \\
& 15 \\
& 45 \\
& \underline{60} \\
& 646.815 \\
+ \text{ remainder:} & \underline{+\ 7.385} \\
& 654.200
\end{array}
$$

In general, when dividing two floating-point numbers, one of the following three conditions will exist:

1. The divisor has no decimal place.
      Example: F8.3 ÷ F3.0
      Position diagram: _ _ _ _ ._ _ _ ÷ _ _ ._
      Result: F8.3   same as the dividend
      Remainder: (1) At most two digits (from divisor) and,
                            (2) three decimal places since the dividend has three

      Example: 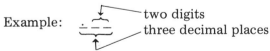 two digits
                                  three decimal places

2. The divisor has decimal places less than or equal to the dividend.
      Example: F5.3 ÷ F3.2
      Position diagram: _ ._ _ _ _ ÷ ._ _ _

      Shift two places:

      _ _ _ _ ._ _ ÷ _ _ ._

Result: F5.1
Remainder: (1) At most two digits since the divisor is **F3.2**
and
(2) three decimal places since the dividend has three

Example: 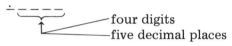   F4.3

three decimal places includes two digits

3. The divisor has decimal places greater than the dividend.
Example: F6.2 ÷ F5.4
Position diagram: _ _ _ . _ _ ÷ . _ _ _ _ _

Shift four places:

_ _ _ _ _ _ _ . ÷ _ _ _ _ .

Result: F8.0
Remainder: . _ _ _ _

four digits

five decimal places

EXERCISES

1. Write the format for the result field.
    (a) F6.3 + F8.4   (b) F5.2 – F4.3
    (c) F6.2 + F7.2 – F5.3   (d) F5.2 × F4.2
    (e) F6.4 × F5.1

2. Identify the format for the result field. Verify your answer by doing the actual calculations.
    (a) 8.25 + 2431.4   (b) .004 + 2.0538 + 1.5786
    (c) 98.73 – 15.1   (d) 17. – 9.8432
    (e) 6.01 × 5.4   (f) 17.132 × 1.05
    (g) (9.15 + 7.092) – 8.53   (h) 7.1 × 9.2 × 11.4
    (i) (8.53 + 10.21) × .055   (j) (.075 × 2050.45) + (.045 × 1450.50)

3. Identify the result field and the remainder field.
    (a) F5.2 ÷ F3.1   (b) F7.2 ÷ F3.2
    (c) F4.1 ÷ F3.2   (d) F6.3 ÷ F3.1
    (e) F5.1 ÷ F4.3

4. Identify the result field and the remainder field. Verify your answer by doing the actual calculations.
    (a) 38.7 ÷ 3.42   (b) 50.89 ÷ 14.5
    (c) 8294.843 ÷ 4.12   (d) 9.84532 ÷ 5.21
    (e) 0.00479421 ÷ 0.004   (f) 583.2 ÷ .421

     (g) $9483.1 \div 67.433$ (h) $87000.1 \div .0003$

     (i) $673. \div .02$ (j) $(9.31 \times .54) \div 2.3$

5. Fill in the blanks with the maximum operator field.

     (a) F7.4 + ——————— = F9.5

     (b) F4.1 + ——————— + F3.2 = F7.3

     (c) F5.2 − ——————— = F7.3

     (d) F4.3 × ———————= F6.4

     (e) ——————— − F3.2 = F6.3

     (f) ——————— × F4.2= F8.5

     (g) F6.2 ÷ ——————— = F6.0   remainder: F3.2

     (h) F5.1 ÷ ——————— = F6.0   remainder: F4.2

     (i) ——————— ÷ F5.4 = F8.1   remainder: F6.5

     (j) ——————— ÷ F5.4 = F8.0   remainder: F5.4

SOLVING PROBLEMS WITH FLOATING-POINT NUMBERS

Once a programmer understands floating-point number fields and result fields, it becomes easier to format the output for a computer printout. Prior to doing the actual computer calculations, the programmer determines the size of the result field. At this point, the exact answer may not be known. But the *size* of the answer (including decimal point and sign) *must fit* the size of the result field.

When a result field is too large, the result will not be wrong. But when a result field is too small, the result will lose significant digits. Therefore, when solving a problem, examine the numbers that are used as operators and make an "educated guess" about the size of the result field.

EXAMPLE 1

    Print a payroll register. There are at most 500 paychecks printed, and no paycheck is greater than $1000. Include the following information for each employee:

      1.   Employee number (six-digit integer)

      2.   Employee name

      3.   Check amount

    Include a final total for the pay period, and, specify formats for (1) employee number, (2) check amount, and (3) total.

Format Fields:

      1.   Employee number: I6.

      2.   Check amount: Maximum possible check is $1000.00, or F7.2.

3. Totals

Maximum check amount is \$1000.00, or F7.2.

Maximum number of employees is 500, or F4.0.

$$1000.00 \times 500. = 500000.00$$

$$F7.2 \times F4.0 = F10.2$$

EXAMPLE 2

Specify the formats needed to calculate the average for the following test scores:

98
75
83
79
90
77
62
88
74
73
80
72

Solution   To calculate the average, add the test scores together and divide the sum by the number of test scores.

1. 12 test scores require an F3.0 field.

2. Each test score requires an F3.0 field.

3. Total of all test scores: F3.0 $\times$ F3.0 = F5.0.

4. Average: F5.0 $\div$ F3.0 = result F5.0   (remainder F3.0).

EXAMPLE 3

Specify the format fields for the following numbers that are printed on the weekly paycheck and stub of the employee.

1. Employee number (six-digit integer)

2. Department number (three-digit integer)

3. Regular hours (40 hours per week)

4. Regular earnings, current (maximum rate is \$25 per hour)

5. Regular earnings, year to date

6. Overtime hours, current (8 hours maximum)

7. Overtime hours, year to date

8. Overtime earnings, current (1.5 of hourly rate)

9. Overtime earnings, year to date

10.   Total earnings, current

11.   Total earnings, year to date

Format Fields:

1.   I6

2.   I3

3.   Hours: F5.2 (this allows for fractional hours worked)

4.   Regular earnings, current:

$$
\begin{array}{rl}
40 & \text{hours maximum} \\
\times\ \$25.00 & \text{per hour maximum} \\
\hline
\$1000.00 & \text{maximum salary, or F7.2}
\end{array}
$$

5.   Regular earnings, year to date:

$$
\begin{array}{rl}
\$1000.00 & \text{maximum salary} \\
\times\ \ 52 & \text{weeks in a year} \\
\hline
\$52000.00 & \text{maximum earnings, or F8.2}
\end{array}
$$

6.   Overtime hours, current: 8.00 hours maximum, or F4.2

7.   Overtime hours, year to date:

$$
\begin{array}{rl}
8.00 & \text{hours per week maximum} \\
\times\ 52 & \text{weeks} \\
\hline
416.00 & \text{maximum hours per year, or F6.2}
\end{array}
$$

8.   Overtime earnings, current:

$$
\begin{array}{rl}
25.00 & \text{maximum hourly rate} \\
\times\ \ 8.00 & \text{maximum hours} \\
\hline
200.00 & \\
\times\ \ 1.5 & \text{overtime rate} \\
\hline
300.00 & \text{maximum weekly overtime earnings, or F6.2}
\end{array}
$$

9.   Overtime earnings, year to date:

$$F6.2 \times F3.0 = F8.2$$

10.   Total earnings, current:
Maximum regular earnings: F8.2
Maximum overtime earnings: F4.2
Result: F9.2

11.   Total earnings, year to date:

$$F9.2 \times F3.0 \ (52 \text{ weeks in a year}) = F11.2$$

EXERCISES

1. Find the operand and result fields for

$$\frac{32100 \times 0.004156}{876420 \times 0.000000123}$$

2. A gallery is marking all of its paintings down by 25%. Paintings range from $12,985 to $75. Specify fields for the operands and results to calculate the sale prices of the paintings.

3. Give the fields used to calculate the total profit for a retailer on 1200 items bought for $2.55 each and marked up 40%.

4. Given the following table, determine the size of the operand field for:
   (a) Model number.
   (b) Tank capacity.
   (c) Miles per gallon.
   (d) Total distance on one tank.
   (e) Cost for fill-up at $1.35 per gallon.

Model	Tank capacity (gal.)	Miles per gallon	Total distance on one tank (miles)	Cost for fill-up at $1.35 per gallon
1234	8	35	_____	$ _____
1334	10	28.5	_____	_____
1356	12	22	_____	_____
2442	16	18.5	_____	_____
2542	22	15.8	_____	_____
3441	28	12.2	_____	_____

5. The water bill for a certain town lists the following information.
   (a) Current meter reading (three digits).
   (b) Previous meter reading.
   (c) Usage (100 cubic feet).
   (d) Current water charge ($1.10 per 100 cubic feet).
   (e) Current sewer (maximum charge is $50.00 per month).
   (f) Amount due.
   Determine the size of the field for each item.

6. The employees at a certain store get a minimum wage of $3.50 per hour, plus a 15% commission on their individual sales. The maximum weekly sales for any employee would be at most $9000. Determine the size of the fields for each of the following.
   (a) Hours worked per week (40 hours maximum).

(b) Hourly wage.
(c) Sales per week.
(d) Commission.
(e) Total earnings for week.

7. Using the paycheck problem in Example 3, determine the size of the fields for each of the following additional items.
   (a) FICA, current (maximum of 10% of total earnings).
   (b) FICA, year to date.
   (c) Federal taxes, current (maximum of 50% of total earnings).
   (d) Federal taxes, year to date.
   (e) State taxes, current ($2\frac{1}{2}$% ot total earnings).
   (f) Total deductions.
   (g) Net pay.

8. The monthly statement for a home mortgage loan lists the following data.
   (a) Loan number (10-digit integer).
   (b) Loan payment.
   (c) Principal, current.
   (d) Principal, balance (amount still due on loan).
   (e) Interest, current.
   (f) Interest, year to date.
   (g) Escrow, current.
   (h) Escrow, year to date.

   The loan was for $50,000 at 11% per year for 30 years. The monthly payment for the loan, the interest, and the escrow is $676.00. The escrow is the amount paid monthly into a separate account, which is used to pay yearly taxes of $2610.00. Determine the size of the fields needed for items (a) to (h).

ROUNDING AND TRUNCATING

Floating-point numbers have a restriction placed on them, in that they have to be finite because computer representation will always be limited to a set number of decimal places.

For transactions that involve money, more than two decimal places would be meaningless. In other situations, such as determining the amount of paint needed, buying .23 of a gallon would be impossible. In these situations, and many others, the programmer must decide whether the number is to be rounded or truncated. In either case, the result must have practical value to the user.

There are two types of numbers to use with a computer: the integer and the floating-point number. The integer is always a whole number. But a floating-point number can be changed to an integer. To do this, the decimal portion of the number is dropped without affecting the value of the whole number. This process is known as *truncation*. For example, to change 5.78 to an integer, write INT(5.78) = 5. The decimal .78 is dropped (or truncated).

Using truncation, any decimal number can be rounded. The number positions in a decimal number are the following:

| Thousands = 10^3 | Hundreds = 10^2 | Tens = 10^1 | Ones = 10^0 | . | Tenths = $\frac{1}{10}$ | Hundredths = $\frac{1}{100}$ | Thousandths = $\frac{1}{1000}$ | Ten thousandths = $\frac{1}{10000}$ |

To round 5.78 to the nearest ones:

1. Divide 5.78 by 1. $5.78 \div 1 = 5.78$
2. Add .5. $5.78 + .5 = 6.28$
3. Truncate by changing to an integer. $INT(6.28) = 6$
4. Multiply by 1. $6 \times 1 = 6$

5.78 rounded to the nearest ones is 6.

To round 758.43 to the nearest tens:

1. Divide 758.43 by 10. $758.43 \div 10 = 75.843$
2. Add .5. $75.843 + .5 = 76.343$
3. Truncate. $INT(76.343) = 76$
4. Multiply by 10. $76 \times 10 = 760$

758.43 rounded to the nearest tens is 760.

In general: To round a number to the nearest N number place, do the following:

1. Divide the number by N.
2. Add .5.
3. Truncate.
4. Multiply by N.

To round, the process is to shift the decimal point to the place where it is to be rounded, and add .5. This ensures a carry for any place having a number of 5 or more. Otherwise, there will be no carry.

Using this process for decimal positions, the decimal point must be shifted to the right. This occurs because dividing by $\frac{1}{10}$ is the equivalent of multiplying by 10. This comes from the rule of dividing fractions:

$$a \div \frac{1}{b} \qquad \text{becomes} \qquad a \times b$$

To round 3.7843 to the nearest hundredth, which is the $\frac{1}{100}$ place:

1. Divide 3.7843 by $\frac{1}{100}$. $3.7843 \div \frac{1}{100}$
     $= 3.7843 \times 100 = 378.43$
2. Add .5. $378.43 + .5 = 378.93$
3. Truncate. $\text{INT}(378.93) = 378$
4. Multiply by $\frac{1}{100}$. $378 \times \frac{1}{100} = 3.78$

3.7843 rounded to the nearest hundredth is 3.78.

EXERCISES

1. Truncate.
   (a) 125.4783   (b) 1.452
   (c) 10001.2   (d) .1487
   (e) .00052
2. Round 18735.4537 to the nearest:
   (a) Hundreds.   (b) Tenths.
   (c) Thousandths.   (d) Ten thousandths.
   (e) Ones.
3. Following are the week's attendence figures for a total enrollment of 1540 students.

Day	Attendance	Percent present
Monday	1523	————————
Tuesday	1508	————————
Wednesday	1470	————————
Thursday	1498	————————
Friday	998	————————

   (a) Calculate the percent present for each day. Percent present = (attendance ÷ 1540) × 100.
   (b) Round the percent to the nearest ones.
   (c) Round the percent to the nearest tenths.
   (d) Round the percent to the nearest hundredths.
   (e) Complete the tables.

Percent to nearest ones		Attendance to nearest whole number
M _____	× 1540 =	_____
T _____	× 1540 =	_____
W _____	× 1540 =	_____
T _____	× 1540 =	_____
F _____	× 1540 =	_____

Percent to nearest tenths		Attendance to nearest whole number
M _____	× 1540 =	_____
T _____	× 1540 =	_____
W _____	× 1540 =	_____
T _____	× 1540 =	_____
F _____	× 1540 =	_____

Percent to nearest hundredths		Attendance to nearest whole number
M _____	× 1540 =	_____
T _____	× 1540 =	_____
W _____	× 1540 =	_____
T _____	× 1540 =	_____
F _____	× 1540 =	_____

(f) Determine to which place the percent should be rounded to get the most accurate results.

ROUNDING FLOATING-POINT NUMBERS AND PROBLEM SOLVING

When solving problems, both the INPUT and the OUTPUT must be specified. The computer calculations take place to produce the OUTPUT.

EXAMPLE

The tax rate is 3.7452 for every $100 of assessed valuation. Calculate the amount of taxes owed on a $70,525 property.

Solution   The values that will be the INPUT and must be specified are:

Tax rate   3.7452   F6.4

| Assessed valuation | 100.00 | F6.2 |
| Property value | 70525.00 | F8.2 |

To calculate the taxes owed:

(property value ÷ assessed valuation) × tax rate

(70525.00 ÷ 100.00) × 3.7452

705.25 × 3.7452 = 2641.30230

The number of decimal places beyond two is really meaningless. The answer field should allow for only two decimal places. In this case

(F8.2 ÷ F6.2) × F6.4

F8.2 × F6.4 = F13.6

Rounding will reduce the decimal places to two. The result will be

F9.2 OUTPUT: F9.2

Rounding should always be done to the answer and not to the operands. The answer to the problem is $2641.30 after rounding. Consider what happens when the tax rate is rounded. First, 3.7452 becomes 3.75. Then

(70525.00 ÷ 100.00) × 3.75 = 2644.6875

The result rounded to two decimal places becomes $2644.69, and is

the difference 2644.69 - 2641.30 = 3.39

EXERCISES

1. Calculate gross pay for the following group of employees.

Employee no.	Hours worked	Rate of pay	Regular earnings	Overtime earnings	Gross pay
1543	30	$4.75	$ _____	$ _____	$ _____
1602	40	4.95	_____	_____	_____
1784	42	5.25	_____	_____	_____
1842	45	6.00	_____	_____	_____
2044	35	4.80	_____	_____	_____

Specify the input fields for:
(a) Employee number.

(b) Hours worked.
(c) Pay rate.
Calculate and specify fields for:
(d) Regular earnings.
(e) Overtime earnings—rate is 1.5 for any hours over 40.
(f) Gross pay.

2. The basic formula for calculating *interest* is

$$\text{principal} \times \text{rate} \times \text{time}$$

The *maturity value* of a bank loan is the principal + interest. Given the following data:

Principal	Rate (%)	Time (days)
$15,000	11	90
7,500	11.5	75
245.50	12	60
1,750.25	13	35
10,500.75	12.5	45

(a) Specify input fields for:
  (1) Principal.
  (2) Rate.
  (3) Time.
(b) Calculate each of the following and round where appropriate. (*Note:* Days are a fraction of a year.)
  (1) Interest.
  (2) Maturity value.
(c) Specify output fields for:
  (1) Interest.
  (2) Maturity value.

3. A gift shop has a 45% markup on all goods it sells. The *selling price* is cost + markup. The markup is cost × 45%. The manager has received the following information.

Item no.	Cost	Selling price
127	$ 4.50	$_____
126	3.75	_____
125	2.90	_____
124	1.85	_____
384	10.90	_____
370	11.75	_____
350	9.80	_____
542	7.85	_____
530	12.50	_____

(a) Specify input fields for:
    (1) Item number.
    (2) Cost.
(b) Calculate cost (round where necessary).
(c) Specify output fields for the selling price.

4. For each of the following, determine (1) input and input fields, (2) output and output fields, and (3) calculations, rounding where required.

(a) A salesman receives a base wage plus commissions. His yearly salary is $10,500 and his commissions are 20% of sales. What was his gross pay for the month when his sales totaled $1250.75?

(b) A property was bought for $93,750.00. To compute the tax bill:

$$\text{Assessed value} \times \frac{\text{board of review}}{\text{equalization factor}} \times \text{Tax rate}$$

Find the assessed value if it is one-third of the market price. Then find the tax owed for the property if the board of review equalization factor is 1.0176 and the tax rate is 8.380%.

(c) A retailer paid $5.76 for 100 craftware items. The items were marked up 45%. If 45 were sold at the original price, 25 were sold at 60% of the original price, and the remainder were sold at 50% of the original price, what was the profit on these items?

SUMMARY

Computers have long since proven their value in number arithmetic by their extremely fast and efficient operations of addition, subtraction, multiplication, and division. The only additional instructions it needs for these operations is to be told if the numbers are floating point or integer, and the formats of the numbers.

By applying the information on floating-point numbers, integers, and formats given in this chapter, the results may be formatted in advance so that the answer will be correct. Truncating and rounding form a part of the process of formatting results to make them meaningful.

CHAPTER REVIEW

1. Fixed-point numbers are the set of numbers also called ——————— .
2. Decimals are known as ——————— numbers.
3. When adding decimal numbers, the numbers are lined up by the ——————— .

4. A field of F6.4 has —— positions for whole numbers and —— positions for decimals.

5. When multiplying two decimal numbers, the total number of places will be the sum of the ——————— , ——————— , and one place for the ———————— .

6. When dividing two decimals, the result field can be as large as the ———————— .

7. The remainder will have at most the same number of digits as in the ———————— .

8. The formula for checking a division result is (———— × ————) + ———————— = ———————— .

9. When a result field for a calculation is too ————, significant digits can be lost.

10. Dropping the decimal portion is called ———————— .

11. To round to the nearest N place:
    (a) ———————— number by N.
    (b) Add —— .
    (c) ———————— .
    (d) ———————— by N.

Solving Problems Using Input, Process, and Output

chapter 4 _____

The computer is a machine that cannot reason. It has nothing in its memory except what has been put there by the program. For example, if the program deals with a "week," the computer does not know that a week has 7 days. If the computer is to have this information, it must come from the program.

In programming, the computer is, of course, only a tool. For this "tool" to do its work properly, every step must be defined and provided for, or the answer will be wrong. The programmer must keep this in mind when approaching every problem.

The simplest way to approach any problem given for computer solution is to ask:

1.   What does the machine need to start?
2.   What is to be done?
3.   What will the machine produce?

This approach relates directly to the basic components of every computer program: *input* (1); *process* (2); and *output* (3).

This chapter will deal primarily with solving problems by using these three basic components of a program. In addition, "algorithms" will be used. *Algorithm* is the name given to a series of steps which are used to solve a problem. These steps are such that they can be applied to any problem with similar elements.

INPUT, PROCESS, AND OUTPUT

The following examples show how the input, process, and output format can be applied to various programming situations that on the surface may seem different, but which actually have similar basic elements.

EXAMPLE 1

Calculate the simple interest owed on a bank loan of $1500 borrowed for 3 years at a rate of 17% per year.

Solution:

Input:	Amount borrowed = $1500.00
	Rate = 17% (or .17)
	Time = 3 years
Process:	Simple interest = amount borrowed $\times$ rate $\times$ time (in years)
	Interest = $1500.00 $\times$ .17 $\times$ 3
	Interest = $765.00
Output:	$765.00 interest owed

EXAMPLE 2

Find the weekly earnings of an employee whose yearly salary is $18,650.

Solution:

Input:	Yearly salary = $18,650.00
	Weeks in a year = 52
Process:	Yearly salary divided by weeks in a year.
	Weekly earnings = 18650 $\div$ 52
	Weekly earnings = 358.65385
	The answer represents a dollar amount; therefore, it must be rounded to two decimal places. For the computer, this is done by using a *rounding algorithm*. It involves (1) multiplying the answer by 100, (2) adding .5, (3) truncating the remainder, and (4) dividing by 100.
	(1) 358.65385 $\times$ 100 = 35865.385
	(2) 35865.385 + .5 = 35865.885
	(3) 35865.885 becomes 35865 (truncated)
	(4) 35865 $\div$ 100 = 358.65
Output:	Weekly earnings = $358.65

EXAMPLE 3

Calculate the weekly salary for an employee who has worked $38\frac{1}{2}$ hours during a week and has an hourly wage of $5.25 per hour.

Solution:

Input:	Hours worked = 38.5 (only decimals are used with the computer)	
	Hourly rate = $5.25	
Process:	Step 1:	Hours worked $\times$ hourly rate
		38.5 $\times$ 5.25 = 202.125
	Step 2:	Round the answer to the nearest cent:
		(202.125 $\times$ 100) + .5 = 20213.0
		20213 $\div$ 100 = 202.13
Output:	Weekly salary = $202.13	

EXERCISES

Do each of the following problems using the input, process, output format. Use the rounding algorithm whenever required.

1. How far can a car travel on 15 gallons of gasoline if it averages 19 miles per gallon?

2. At what rate (miles per hour) is a car going if it travels 350 miles in 8 hours?

3. What is 1 year's interest on $1500 invested for 1 year compounded semiannually at 12%?

4. How much has an item increased in price if the original price was $69 and it was marked up to $75?

5. What is the monthly rental for an office of 2400 square feet if the rental cost is $15 per square foot?

6. Find the rate of travel (miles per hour) for a bike rider who covers a total of 120 miles in 5 hours.

7. Find the earned interest on a 30-day certificate of $50,000 at 22%. (*Hint:* Time must be in terms of years.)

8. A stock originally bought for $4.50 per share is now selling for $7.25 per share. If 400 shares of stock were purchased, what is the total profit and what is the percentage of return?

9. Sales tax is 6%. How much tax is owed on the following purchases: suit, $135.00; blouse, $21.00; stockings, $2.50; and shoes, $55.00?

10. A car averages 24 miles per gallon and travels 5000 miles per year. If gasoline costs $1.23 per gallon, how much does gasoline cost for 1 year of driving?

INPUT, PROCESS, AND OUTPUT WITH DECISIONS

Decisions occur in problems where some specific condition must be recognized. This is done with a comparison function and involves a separate calculation for either a "yes" or a "no" response. The following examples show how the decision (comparison/calculation) function is applied.

EXAMPLE 1

A keypunch operator is paid $180 per week for a regular 40-hour week. What is the operator's gross wages for a 50-hour week if he or she is paid time-and-a-half for all hours over 40?

Solution:

Input:	Weekly earnings = $180 (for a 40-hour week)
	Hours worked = 50
Process:	Are hours greater than ($>$) 40? (comparison function)

        No: Weekly earnings = $180.00
      Yes: Calculate hourly rate: $180 \div 40 = \$4.50$
                Calculate overtime hours: $50 - 40 = 10$
                Calculate overtime wages: $10 \times 4.50 \times 1.5 = 67.50$
                Weekly earnings: $\$180 + \$67.50 = \$247.50$

    Output: Weekly earnings = $247.50

The following example shows another type of decision which involves comparing the results of two calculations.

EXAMPLE 2

A salesperson has two salary options: to work on commision, which is 15% of total sales plus a minimum salary of $2.75 per hour; or to work without commission for a salary of $17,500 per year. If the salesperson can expect minimum sales of $1850 in a 40-hour week, which option will give the salesperson the higher income?

Solution:

Input:	Option 1:	Hours worked per week: 40
		Hourly rate: $2.75
		Commission rate: .15
		Total sales are greater than or equal to ($\geqslant$): $1850.00
	Option 2:	Yearly salary = $17500
		Weeks in a year = 52
Process:	Option 1:	Minimum salary = hourly rate $\times$ hours
		$2.75 \times 40 = 110$
		Commission = total sales $\times$ commission rate
		$1850 \times .15 = 277.5$
		Weekly gross: $110 + 277.5 = \$387.50$

Option 2:   Weekly gross = yearly salary divided by weeks in a year
             17500 ÷ 52 = $336.54
Is option 1 greater than option 2?
    Yes: Choose hourly rate plus commission.
    No: Choose annual salary.
Output: Chosen option

EXERCISES

The following problems involve the use of decisions.

1. A keypunch operator earns $4.50 an hour for a 40-hour week and time-and-a-half for overtime. What is the gross pay for a week of 43 hours?

2. If a person's earnings exceed $2500, then, according to the IRS rules, that person owes federal taxes. Does an employee who earns $125 per week and works 15 weeks owe any taxes?

3. In a particular company, the commission rate is 10% for sales up to $50,000 and 12% for sales over $50,000. What is the total combined commission rate for a salesperson who makes sales of $26,000, $17,000, $19,000, $95,000, and $7000?

4. Students with a grade-point average of 3.5 or greater make the dean's list. The grade points are A, 4; B, 3; C, 2; D, 1; and F, 0. A student's last report card was A, A, B, C, A. Has the student made the dean's list?

5. An employee made $220.58, $210.65, $226.75, $201.90, and $230.80 in the last 5 weeks. Would a yearly salary of $12,000 give a higher average weekly wage?

INPUT, PROCESS, AND OUTPUT WITH COUNTERS

Another type of decision is one that involves a *counter* and a question to determine if the *end of counter* or final value has been reached. A counter has three parts: (1) the *initial value*, from which the counter begins; (2) the *increment value*, which is the amount by which the counter will be increased; and (3) the *final value* of the counter, which is the test to check for "end of counter."

A good example of this type of decision is found in calculations for compound interest. Compound interest problems start with a *simple interest* calculation:

$$\text{interest} = \text{principal} \times \text{rate}$$

The interest is then added to the principal. This amount (which is principal plus interest) is now the principal for the second year's calculation. This process is repeated for the number of years involved in the calculation.

For example, given a principal that is to be compounded yearly for 3 years, the step-by-step solution would be the following:

Year 1:   Interest = principal × rate
            Principal (at the end of the first year) = principal + interest
Year 2:   Interest = principal × rate
            Principal (at the end of the second year) = principal + interest
Year 3:   Interest = principal × rate
            Principal (at the end of the third year) = principal + interest

This step-by-step solution shows that:

1. The year is a "counter" since it starts at 1, is incremented by 1, and the "end of counter" is 3.
2. The same process is repeated each year to find the new interest and principal.

This step-by-step solution can be simplified by incorporating a counter. To do this, the following steps have to be included:

Start: Year counter at 1
Increment: Year counter by 1
Question: Is your counter greater than time? (In this case, 3 years.)
                  The question has two responses: Yes and No.
                  No: means the steps for calculating interest and ending balance have to be repeated (the end of the 3-year period has not yet been reached).
                  Yes: means the calculations are complete (the end of the 3-year period has been reached). The answer has to be rounded to two decimal places and given as "output" or "ending balance."

To prepare a solution algorithm for this kind of calculation, by definition it must work for any number of years. With this in mind, the following approach can be used:

Input: Principal
                  Rate
                  Time length (needed for year counter)

Process: Step 1:   Set year counter to 1

Step 2:   Calculate: interest = principal × rate

Step 3:   Principal = principal + interest

(The quantity on the right is now assigned to the quantity on the left.)

The new amount is the principal at the end of the year.

Step 4:   Increment year counter by 1

Step 5:   Test: Is year counter greater than time length?

No: Return to step 2 and continue.

Yes: Round answer (using the rounding algorithm) to the nearest hundredth and go to "output."

Output: Ending balance = principal

Now apply the same solution algorithm to a specific problem.

EXAMPLE

Find the total amount in an account with a beginning balance of $1000 at an annual rate of 12% compounded yearly for 4 years. (The answer will show the account status at the end of the fourth year.)

Solution:

Input: Principal = $1,000

Rate = 12%

Time = 4 years

Process: Step 1:   Year counter = 1

Step 2:   Interest = 1000 × .12 = 120

Step 3:   Principal (at the end of the year) = 1000 + 120 = 1120

Step 4:   Year counter = 2

Step 5:   Is the year counter greater than 4?

No:

Step 2:   Interest = 1120 × .12 = 134.4

Step 3:   Principal (at end of year) = 1120 + 134.4 = 1254.4

Step 4:   Year counter = 3

Step 5:   Is the year counter greater than 4?

No:

Step 2:   Interest = 1254.4 × .12 = 150.528

Step 3:   Principal (at the end of year) = 1254.4 + 150.528 = 1404.928

Step 4:   Year counter = 4

Step 5:   Is the year counter greater than 4?

No:

Step 2:   Interest = 1404.928 × .12 = 168.59136

Step 3:   Principal (at end of year) = 1404.928 + 168.59136 = 1573.5194

Step 4:   Year counter = 5

Step 5:   Is the year counter greater than 4?

Yes:

(1573.5194 × 100) + .5 = 157352.44

Truncate and ÷ 100

157352 ÷ 100 = $1573.52

Output: Ending balance: $1573.52

EXERCISES

1. With inflation, real estate values are said to increase 1% per month. An apartment building was originally worth $150,000. How much will it be worth in 2 years?

2. A retailer has bought several cases of imported wine. As the wine ages, its value increases 10% each year. It reaches its prime in 10 years. If the retailer originally spent $4806 on the cases, how much is the wine worth when it reaches its prime?

3. Assume that a worker's starting salary is $16,000 a year and that the person will get a raise of $900 a year every year. How much will this person's salary be at the end of 10 years?

4. An oil well yields 400 barrels of crude oil a month and will run dry in 3 years. The price of crude oil is currently $18 per barrel but rises at the rate of 3 cents per month. What is the total revenue from the well?

There is a quantity in mathematics identified as "*n*!" (*n* factorial). It means sequential multiplying up to the number shown. For example, 5! would mean 1 × 2 × 3 × 4 × 5. Develop a counting algorithm to solve the following (but do not solve).

5. 10!

6. 25!

7. 35! – 15!

COUNTERS WITH INCREMENTS OTHER THAN 1

Counters do not need an increment of 1 each time. For example, when interest is compounded semiannually for 3 years, the counter becomes:

Initial value = 1

Increment = .5 (for half-year increments)

Test greater than ($>$) 3? (Is the time greater than 3 years?)

Interest compounded quarterly for 5 years has the following counter:

Initial value = 1

Increment = .25 (for quarter-year increments)

Test $>$5 (Is the time greater than 5 years?)

In problems such as compound interest, increments can be numbers other than 1. The increment is expressed as a decimal: .5 for one-half, .25 for one-fourth, and so on. This is the formula: interest = principal $\times$ rate $\times$ time (in years). The time would be expressed as .5, .25, and so on.

EXERCISES

Exercises 1 to 8 are sequences of numbers that can be generated by a counter. Give the three parts of a counter for each. Find the initial value, increment value, and final value for each.

1. $2, 3, 4, 5, \ldots, 100$
2. $4, 8, 12, 16, 20, \ldots, 200$
3. $3, 4, 5, 6, \ldots, 150$
4. $4, 9, 14, 19, \ldots, 229$
5. $3, 6, 9, 12, \ldots, 123$
6. $7, 10, 13, 16, \ldots, 130$
7. $4, 10, 16, 22, \ldots, 136$
8. $5, 28, 51, \ldots, 442$
9. An object dropped from a building 1200 feet high falls 16 feet during the first second, 48 feet during the second second, and 80 feet during the third second. If the object continues to fall in the same manner, how long will it take to reach the bottom?

Develop only the solution steps for Exercises 10 to 13. Do not solve.

10. Find the final balance in an account with an opening balance of $2500 at a rate of $7\frac{1}{2}$% compounded semiannually for 5 years.
11. Find the final balance for Exercise 10 if it is compounded quarterly.
12. A car depreciates at a rate of $500 per year. If the original cost was $12,000, develop the solution steps to show when the car is worth $6000.
13. A rubber ball dropped from a height of 100 feet bounces .375 of the distance from which it had fallen. Find how many times it will bounce before it stops.

CALCULATIONS WITH A COUNTER

In the examples shown thus far, the counter has had three well-defined parts (initial value, increment value, and final value). However, if the final value is missing, the problem can still be solved.

EXAMPLE

How long will it take for $500 to double if it is invested at 17% compounded annually?

Solution   The initial value of the counter is set at 1, incremented by 1, but instead of a test for "end of counter," substitute a test for balance >$1000. Here the final value of the counter answers the question. The step-by-step solution is as follows:

Input: Principal = 500
           Rate = .17
           Counter set at 1

Process: Step 1:   Interest = principal × rate
           Step 2:   Principal (at the end of the year) = principal + interest
           Step 3:   Is principal (at the end of the year) >1000?
                     Yes: Go to output
                     No: Increment counter by 1
                              Return to step 1 and continue.

Output: Years to double = counter

EXERCISES

The following exercises involve calculations with counters.

1. The population of a metropolitan area is 100,000 and increases at a rate of 4% each year. In how many years will the population double?

2. A bus line carries 5000 passengers per day at a rate of 50 cents per passenger. Use a counter to represent the rate increases and show the solution steps to fill in the following table. If for each 5 cent rate increase the bus line loses 100 passengers, decide when revenues are at a maximum.

Rate	Number of passengers	Bus line revenues

3. A retailer received a shipment of 12,000 one-pound cans of tomatoes which are placed in storage. Every week, 300 cans are sold. The cost of storage is 2 cents per pound per week. What is the total storage paid on the tomatoes?

4. A bus company charters buses to groups of 40 or more people. For

groups of 40, the total charge is $80. For groups greater than 40, the charge is reduced by 50 cents per person for each person in excess of 40. A group of what size will generate the maximum profit?

5. Water is filling a 31,400-cubic foot tank at a rate of 25 cubic feet per minute. How long will it take to fill the tank?

USING COUNTERS TO PRODUCE TABLES

The counter can also be applied to other types of problems, such as depreciation. But for depreciation, the counter will be used to produce a table instead of a final value. In accounting, the straight-line depreciation method involves the original cost, the recycle value, and the useful life in years. The amount of depreciation is the cost minus the recycle value, divided by the number of useful years.

EXAMPLE

A car costing $6000 has a useful life of 10 years. It can be sold for recycling for $200 at the end of 10 years.

Solution   To construct a depreciation schedule for 10 years, the steps would be as follows:

Input: Original cost = $6000
            Useful life = 10 years
            Recycle value = $200

Process: Step 1:   Set up column headings:

Year	Depreciation amount	Total depreciation amount	Value at end of year

         Step 2:   Calculate the depreciation:
                   Depreciation = (6000 − 200) ÷ 10 = 580 per year
         Step 3:   Set counter to 1
         Step 4:   Calculate total depreciation amount = year counter ✕ depreciation
         Step 5:   Calculate value at end of year = original value − total depreciation amount
         Step 6:   Fill in line of table
         Step 7:   Increment year counter by 1
         Step 8:   Is year counter greater than useful-life years?
                 Yes: Finished
                 No: Return to step 4 and continue

Output: Column headings and table

EXAMPLE OF TABLE

At end of year:	Depreciation amount	Total depreciation amount	Value at end of year
1	$580	$ 580	$5420
2	580	1160	4840
3	580	1740	4260
4	580	2320	3680
5	580	2900	3100
6	580	3480	2520
7	580	4060	1940
8	580	4640	1360
9	580	5220	780
10	580	5800	200[a]

[a]$200 is the recycle value.

EXERCISES

1. On a ship, time is represented by striking bells on the hour and half-hour. 12:30 A.M. is identified by one bell, 1:00 A.M. by two bells, 1:30 A.M. by three bells, and so on. There are a maximum of eight bells. After eight bells, the sequence begins again. Develop a solution in which a counter generates a table showing the time and its bell equivalent. Use the following headings.

Time (hours)	Number of bells to strike

2. Write the steps needed to produce a mortgage repayment table for 1 year (12 payments). The amount borrowed is $35,000 at a rate of $8\frac{3}{4}\%$. The monthly mortgage payment is $260.

   To calculate the amount of decrease on a loan, find the interest owed for the month.

   Interest = (principal $\times$ rate) $\div$ 12

   Reduce amount paid by the interest

   Amount paid on loan = monthly amount − interest

   New principal = principal − amount paid on loan

Use the following headings:

Month	Old loan amount	Monthly amount	Amount paid on loan	New loan amount

3. Draw a depreciation table for the following. (Use the depreciation table format given in the example in this section.)

Item	Original cost	Useful life	Recycle value
Machine 1	$6000	10	$400
Machine 2	8500	8	300
Machine 3	7000	11	400
Machine 4	4800	9	200

4. A contest winner received $2000 quarterly. For the past 3 years, the amount has been deposited in an account at a rate of $5\frac{1}{2}\%$, compounded quarterly. Draw a table to show deposits and principal amounts for each year.

5. Construct a table for the following problem. A retailer received a shipment of 15,000 one-pound cans of vegetables which are placed in storage. Every week, 350 cans are sold. The cost of storage is 2.5 cents per pound per week. What is the total storage paid on the vegetables? Use the following headings:

Amount in storage	Week	Storage cost	Total Storage

SUMMARY

A basic problem in programming is "where to start?" This section has answered that question by using a three-part "starting place": input, process, and output. Of these three parts, process is where the solution algorithm belongs.

Very seldom is a process a one-step calculation. As this chapter has shown, the process normally involves several steps in a logical sequence. These steps can include decisions where a question is posed and two or more items are compared. The answer to the question decides which process to use.

The steps in a process can also include counters, which are made up of three parts: initial value, increment value, and question (final value), to decide when the counter (or loop) is finished. These three parts form a counter for repetitious calculations. Variations with counters have also been shown, such as situations where the increment is other than 1, where calculations are involved, and where tables are to be produced.

The three-part "starting place" as well as techniques involving decisions and counters will apply to a wide variety of programming problems found in the working world, and will provide you with a solid foundation on which to build more advanced programming skills.

CHAPTER REVIEW

1. The first step in finding a solution to a programming problem is to find the three components of a program: —————— , —————— , and —————— .
2. The logical steps needed to solve a problem are called a(n) —————— .
3. —————— refers to the information needed to start solving the problem.
4. The solution steps are the —————— .
5. The result of the process is the —————— .
6. If the program is to calculate monthly earnings given a yearly salary, the input must include —————— .
7. A decision involves a —————— which is a question.
8. A decision has two branches which are identified by a ——— or ——— response.
9. Repetitious calculations are simplified by the use of a ——— .
10. A counter has three parts: —————— , —————— , and —————— .
11. The final value of a counter identifies the ——— of counter.
12. To drop the decimal part of a number is to —————— .
13. To use the rounding algorithm to round to the nearest tenth involves: (1) multiplying by ——— , (2) adding ——— , (3) —————— , and (4) (4) dividing by ——— .
14. Given the sequence: 5, 21, 37, . . . , 165, the initial value is ——— , the increment value is ——— , and the final value is ——— .

Algorithms

Computer programming is actually problem solving. To solve a problem the computer merely follows the steps the programmer sets up to solve the problem.

The first step a programmer must take is to understand the problem, that is, to define the three parts of a program: input, process, and output. The *input* is what is given and the *output* is the required information.

The *process* will be the steps used to solve the problem. These steps are the algorithm. An algorithm is the procedure that must be followed to produce the desired results. Algorithms should be:

1. *Precise.* Nothing can be assumed. Every term and step must be defined.
2. *Clear.* Nothing should be ambiguous. All terms and steps must be understood in only one way.
3. *Specific.* Produce the desired output.
4. *Finite.* Produce the output in a limited number of steps.

STRAIGHT-LINE ALGORITHMS

To develop an algorithm, it is necessary to identify the process and to define that process as a step-by-step method. The algorithm is made up of these steps. Certain algorithms are called *straight line*; they proceed from step 1 to step 2 to step 3, and so on, until the final step of the solution is reached.

EXAMPLE 1

Calculate the average of three exams.

Solution:

Input:	Three exams
Output:	The average
Process:	Calculate the average
Algorithm:	Step 1: Exam 1 + Exam 2 + Exam 3 = SUM
	Step 2: SUM ÷ 3 = Average

EXAMPLE 2

Calculate the net pay of each of the following:

Employee	Hourly rate	Hours worked	Net pay
A	$3.85	37	$_____
B	5.00	38	_____
C	4.95	40	_____
D	7.85	39	_____
E	9.52	36	_____
F	8.75	38	_____

Solution:

Input:	Hourly rate
	Hours worked
Output:	Net pay
Process:	Calculate net pay
Algorithm:	Step 1: Hourly rate × hours worked = net pay

The algorithm is stated in such a way that it will always calculate net pay for the various hourly rates and the hours worked.

EXAMPLE 3

Fill in the following table.

Car	EPA rating (miles per gallon)	Speed (miles per hour)	Consumption (gallons per minute)
Alfa Romeo	17	55	_____
BMW	18	55	_____
Cadillac	26	55	_____
Corvette	15	55	_____
Datsun	33	55	_____
De Lorean	19	55	_____
Honda	34	55	_____
Horizon	30	55	_____
Toyota	39	55	_____

Solution:

Input: Miles per gallon (mpg)
         Speed in miles per hour

Output: Consumption in gallons per minute

Process: Calculate gallons per hour and change to gallons per minute.

Note: Mpg is $\dfrac{\text{miles}}{\text{gallons}}$

Speed is $\dfrac{\text{miles}}{\text{hour}}$

$$\frac{\cancel{\text{miles}}}{\text{hour}} \times \frac{\text{gallons}}{\cancel{\text{miles}}} = \frac{\text{gallons}}{\text{hour}}$$

$$\text{speed} \times \frac{1}{\text{mpg}} = \frac{\text{speed}}{\text{mpg}}$$

The problem is: speed ÷ mpg

Algorithm: Step 1:   Speed ÷ mpg $= \dfrac{\text{gallons}}{\text{hour}}$

Step 2: $\dfrac{\text{gallons}}{\text{hour}} \div 60 = \dfrac{\text{gallons}}{\text{minute}}$

EXERCISES

1. Find the solution.

    Input: Seconds = 13872
    Process: Step 1:   Divide seconds by 60
                            Minutes = dividend
                            Seconds = remainder
                  Step 2:   Divide minutes by 60
                            Hour = dividend
                            Minutes = remainder
    Output: Hour, minutes, seconds

2. Find the solution.

    Input: Kilometers = 5
    Process: Step 1:   Feet = kilometers $\times$ 3281
                  Step 2:   Miles = feet $\div$ 5280
                  Step 3:   Yards = feet $\div$ 3
                  Step 4:   Inches = yards $\times$ 36
    Output: Kilometers, miles, yards, feet, inches

3. Develop an algorithm for each problem.
    (a) Determine the amount of change for a given purchase.
    (b) Calculate the cost of sale items reduced 15%.
    (c) Determine the percent of goal reached in a charity campaign.
    (d) Change miles into inches (1 mile = 5280 feet).
    (e) Determine the final amount owed on a purchase for a sale mark-down of 20% and a 5% discount for paying cash.

4. Determine the algorithms needed to fill in the tables.

    (a)

Stock	Purchase price	Current price	Profit or (loss)
A	$55.75	$60.00	$ _____
B	14.25	12.80	_____
C	9.50	9.75	_____
D	44.00	45.75	_____
E	56.75	65.25	_____

    (b)

		COST		
Per year	Per week	Per day	Per hour	Per minute
$85,000				
12,500				
60,000				

(c)   RATE OF RETURN ON AN INVESTMENT

Price per share	Number of shares	Yearly dividends	Percentage of return
$1000.	3	$226.20	_____
8.75	50	100.00	_____
16.50	100	212.00	_____
20.25	100	280.00	_____

ALGORITHMS WITH REPEATING STEPS

Certain calculations involve repeating steps: for example, 6! (factorial). 6 factorial is $1 \times 2 \times 3 \times 4 \times 5 \times 6$.

EXAMPLE 1

Calculate 11!

Solution algorithm:

Input: $N = 11$

Process: Step 1: $I = 1$
                      Product = 1
            Step 2:   Product = product $\times$ I
            Step 3: $I = I + 1$
            Step 4:   If $I \leqslant N$, go to step 2
            Step 5:   Go to "output"

Output: Product

One type of problem that involves repeating steps is the calculation of an average. The repetition will involve the use of a counter.

EXAMPLE 2

Calculate the average of 50 numbers.

Solution algorithm:

Input: $N = 50$

Process: Step 1:   Sum = 0
            Step 2:   Counter = 1
            Step 3:   Input number
            Step 4:   Sum = sum + number
            Step 5:   Counter = counter + 1
            Step 6:   If counter $\leqslant N$, go to step 3
            Step 7:   Average = sum $\div N$
            Step 8:   Go to "output"

Output: Average

By using N in the algorithm rather than 50, it can be used for any number. This number will be specified in the input but the algorithm will remain the same. An algorithm must work for all problems of a given type.

EXAMPLE 3

In mathematics, a sum such as $1 + 3 + 5 + 7 + 9 + 11$ can be shortened to

$$\sum_{I=1}^{6} 2I - 1$$

$2I - 1$ is the formula for odd numbers. The summation is initialized at $I = 1$ and terminated at 6. By taking I from 1 to 6 and evaluating $2I - 1$, the problem is

I	$2I - 1$
1	$2 \times 1 - 1 = 1$
2	$2 \times 2 - 1 = 3$
3	$2 \times 3 - 1 = 5$
4	$2 \times 4 - 1 = 7$
5	$2 \times 5 - 1 = 9$
6	$2 \times 6 - 1 = 11$

Solution algorithm:

Input:	Final value = 6	
Process:	Step 1:	Sum = 0
	Step 2:	Counter = 1
	Step 3:	Number = 2 × counter - 1
	Step 4:	Sum = sum + number
	Step 5:	Counter = counter + 1
	Step 6:	If counter ⩽ final value, go to step 3
	Step 7:	Go to "output"
Output:	Sum	

Another kind of problem using a counter would involve some type of compounding.

EXAMPLE 4

How much is in an account of $1000 compounded yearly for 10 years at 8%?

Solution algorithm:

Input:	Amount = 1000
	Rate = .08
	Time = 10
	Year = 1

Process: Step 1: Counter = 1
          Step 2: Interest = amount × rate × year
          Step 3: Amount = amount + interest
          Step 4: Counter = counter + 1
          Step 5: If counter ≤ time, go to step 2
          Step 6: Go to "output"
Output: Amount

The same algorithm can be applied to problems involving compounding semi-annually, monthly, weekly, and so on. In these cases, the time would be adjusted. If this problem were to be compounded semiannually, it would have times of 20, since there are 20 half-years in 10 years and year = $\frac{1}{2}$.

In general, when an algorithm is developed for a certain kind of problem, it must apply to other problems of the same type.

EXAMPLE 5

The population of a small city was 120,000 in 1980. Predict the population in 1990 if the population:

(a) Increases 6% yearly.

(b) Increases 3% yearly.

(c) Decreases 1% yearly.

Solution algorithm An algorithm to solve these three problems (and other similar problems) would be:

Step 1: Year = 1

Step 2: Change = rate × population

Step 3: Population = population + change

Step 4: Year = year + 1

Step 5: If year ≤ 10, go to step 2

Step 6: Go to "output"

The one algorithm will apply to all three parts of the problem. The only differences will be the input.

For part (a), the input will be: rate = .06.

For part (b), the input will be: rate = .03.

For part (c), the input will be: rate = −.01.

In the case of part (c) the algorithm will work the same. However, since the rate is negative (because the population is decreasing), the change will be negative and the population will be decreasing by the change.

The algorithm should always be written to apply to a general type of problem. Use only names or variables. Do not use a specific number since this decreases the applications for the specific algorithm.

EXERCISES

1. Solve, using the algorithms in this section.
   (a) 8!
   (b) Calculate the average of 95, 78, 90, 83, 69.
   (c) How much is an account with a beginning balance of $500, compounded semiannually for 4 years at $5\frac{1}{2}\%$?

2. Develop an algorithm to solve each problem.

   (a) $\displaystyle\sum_{I=1}^{6} 2I - 5$ (b) $\displaystyle\sum_{J=5}^{10} 3J$

   (c) $\displaystyle\sum_{K=1}^{5} 10 - 2K$ (d) $\displaystyle\sum_{K=3}^{6} K^2 - 3K$

   (e) $\displaystyle\sum_{J=5}^{11} \frac{J}{2}$

3. Develop an algorithm to find the population of a certain city in 1985 if the population growth is 6% per year and the population in 1980 was 250,000.

4. The Fibonacci sequence is 1, 1, 2, 3, 5, 8, 13, 21, and so on.

$$F_1 = 1$$
$$F_2 = 1$$
$$F_3 = F_1 + F_2$$
$$F_4 = F_2 + F_3$$
$$\vdots$$
$$F_{N+1} = F_{N-1} + F_N$$

Develop an algorithm to find the fifteenth number in the sequence.

5. Expand your algorithm for compound interest. The original deposit was $100. Each month another $100 was deposited. If the rate was 6% compounded monthly, how much was in the account after 5 years?

6. Develop one algorithm that can be applied to solve the following. Which investment gives the best return on $10,000 at the end of 10 years?

(a) Compounded yearly at 14%.
(b) Compounded monthly at 10%.
(c) Compounded quarterly at 12%.

ALGORITHMS WITH DECISIONS

Some problems involve a decision that will yield different solutions depending on the yes or no decision answer.

EXAMPLE 1

The problem of net pay will have two different solutions if the problem involves overtime. Overtime will be calculated for hours over 40 for the week. The question will be "Are hours > 40?"

If *no*, then net pay = hours worked $\times$ rate

If *yes*, then overtime = (hours worked - 40) $\times$ 1.5 $\times$ rate

Regular pay = 40 $\times$ rate

Net pay = overtime + regular pay

Solution algorithm   An algorithm in this case must provide for both the yes and no solutions.

Input:	Hours worked	
	Rate	
Process:	Step 1:	Overtime = 0
	Step 2:	Are hours worked > 40?
		Yes: Overtime = (hours worked - 40) $\times$ 1.5 $\times$ rate
		Hours worked = 40
		No: Go to step 3
	Step 3:	Regular pay = hours worked $\times$ rate
	Step 4:	Net pay = regular pay + overtime
	Step 5:	Go to "output"
Output:	Net pay	

By setting overtime to 0, step 4 in an instance where hours are $\leqslant 40$ will have a 0 added to regular pay. In this case regular pay and net pay are equal. When hours are > 40, overtime will be calculated and hours worked will be reset to 40.

EXAMPLE 2

If the population of a certain city of 200,000 people increases at the rate of 3% yearly, when will it reach 250,000?

Solution algorithm   Here the algorithm must involve a question about the result of a calculation.

Input: Population = 200,000
            Rate = .03
            Increased population = 250,000

Process: Step 1:   Year = 1
            Step 2:   Increase = rate × population
            Step 3:   Population = population + increase
            Step 4:   Is population ≥ increased population?
                         Yes: Go to "output"
                         No: Go to step 5
            Step 5:   Year = year + 1
            Step 6:   Go to step 2

Output: Year

Certain problems can have a series of questions.

EXAMPLE 3

The commission rate for salespersons is according to the following table:

Sales	Rate
Up to $10,000	8%
Next $5000	10%
Next $5000	12%
All sales over $20,000	15%

Solution algorithm:

Input: Sales

Process: Step 1:   Sales > $10,000?
                         No: Commission = .08 × sales, go to "output"
                         Yes: Sales = sales − $10,000
                                Commission = $10,000 × .08
            Step 2:   Sales > $5,000?
                         No: Commission = commission + sales × .10; go to
                                "output"
                         Yes: Sales = sales − $5000
                                Commission = commission + $5000 × .10
            Step 3:   Sales > $5000?
                         No: Commission = commission + sales × .12; go to
                                "output"
                         Yes: Sales = sales − $5000
                                Commission = commission + $5000 × .12
            Step 4:   Commission = commission + sales × .15
            Step 5:   Go to "output"

Output: Commission

EXERCISES

1. Use the overtime algorithm to complete the table.

Hours Worked	Rate	Net Pay
38	$6.50	$ _____
42	7.80	_____
45	8.50	_____
36	9.00	_____
50	9.25	_____

2. Use the commission algorithm to complete the table.

Sales	Commission
$ 9,500	
12,750	
22,000	
18,000	
20,000	

3. Develop an algorithm for the following. In how many years will an investment of $500 double if it is invested at $5\frac{1}{4}\%$ annually?

4. Develop an algorithm for calculating postage on first-class mail. The rate is 20 cents for up to the first ounce and 17 cents for every ounce thereafter.

5. Use the algorithm you developed in Exercise 4 to complete the table.

Weight (ounces)	Postage (cents)
2	_____
4	_____
$\frac{1}{2}$	_____
$2\frac{1}{2}$	_____

6. Develop an algorithm for calculating cost given the following information.

Thermal Unit	Rate per Thermal Unit (cents)
Up to 50	35
Next 30	33
Next 20	31
Over 100	28

7. Develop an algorithm to determine the cost recovery for a company when it purchases a new machine. The machine costs $25,000, and for every 2000 units produced the company saves $200. How many units should be produced to pay for the machine?

8. In a company, the manual cost of production is $25 per 10 units. The computerized cost of production is $8 per 10 units, with an initial expenditure of $15,000. *Develop an algorithm* to determine how many units must be produced by the computerized operation before the company can receive a profit.

9. Use the algorithm you developed in Exercise 8 *to solve* that problem.

SUMMARY

The algorithm must be:

1. Precise—every term defined
2. Clear—all steps understood
3. Specific—produce the specified results
4. Finite—limited number of steps

Once the three parts of a problem are identified (input, process, output), the next step is to define the algorithm. The algorithm specifies the steps of the process.

Algorithms developed for a specific problem should be defined so that they fit all problems of the same type. This is done by using general terms rather than the specifics of the problem in the algorithm.

CHAPTER REVIEW

1. The three parts of a program are _____ , _____ , and _____ .

2. The input is the _____ information and the output is the _____ information.

3. An algorithm is a set of _____ that solve the problem.

4. A repeating algorithm must have a _____ to determine when it should end.

5. $\sum\limits_{J=1}^{5} 3J - 2$ is a _____ of _____ terms.

6. Compounding quarterly for 5 years means that a counter is executed _____ times.

7. Problems that involve questions usually have two branches: one for _____ and the other for _____ .

8. A repeating algorithm must have _____ because algorithms must be finite.

Flowcharts

The flowchart is the pictorial representation of an algorithm. A flowchart is a way of visually seeing how the algorithm works to produce the desired result. By tracing the flowchart, the algorithm can be checked to determine if it will work correctly.

This chapter will deal only with flowcharting as it applies to numerical problems.

BASIC SYMBOLS

Problem solving with a computer involves the three parts of input, process, and output. These parts are represented in flowcharting by the following symbols:

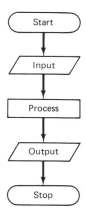

Because an algorithm is finite, the flowchart must show where the algorithm finishes with a *Stop* symbol.

The following is a flowchart to calculate the area of a rectangle:

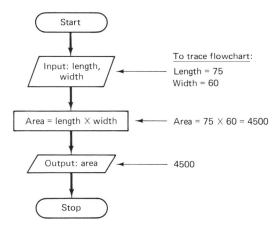

The shape of the boxes in a flowchart represents functions, and the functions should match the algorithm. If, for example, the problem is to flowchart an algorithm for changing seconds into hours, minutes, and seconds, it would be written as follows:

Input: Seconds

Process: Step 1:   Minutes = Seconds ÷ 60
               Step 2:   Seconds = Remainder
               Step 3:   Hours = Minutes ÷ 60
               Step 4:   Minute = Remainder
               Step 5:   Go to "Output"

Output: Hours, Minutes, Seconds

To draw the flowchart, the input and output will each have their own boxes. The process, which is the algorithm, will have *one box for each step*.

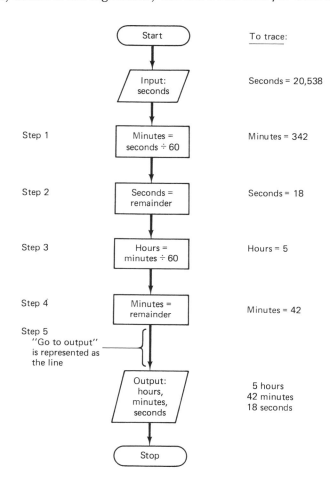

EXERCISES

Match statements 1 to 5 with the appropriate flowchart symbol.

1. Start
2. Output
3. Input
4. Process
5. Go to
6. Write the algorithm for the flowchart.

A.

B.

C.

D.

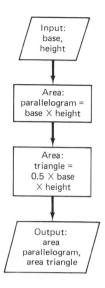

7. Write the algorithm for the flowchart.

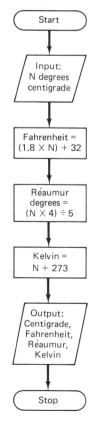

8. Trace the flowchart in Exercise 6 for base = 115 inches and height = 92 inches.

9. Trace the flowchart in Exercise 7 using 80 degrees centigrade.

10. Draw the flowchart for the following algorithm to calculate state income taxes.

Input: Gross income
              Number of exemptions

Process: Step 1:   Deductions = Number of exemptions $\times$ \$1000
              Step 2:   Taxable income = Gross income – Deductions
              Step 3:   Taxes = Taxable income $\times$ .025
              Step 4:   Go to Output

Output: Gross income, Taxes

FLOWCHARTS WITH DECISIONS

In addition to the basic flowchart symbols shown earlier, the flowchart symbol for a question, decision, or test is a diamond:

Each box used for the earlier symbols had only one exit. However, the diamond symbol can provide for multiple exits. For example, assume that a certain step in an algorithm is a question. The answer to that question will determine the next step. The diamond symbol shows this with "yes" or "no" exits:

The following is an algorithm with a test that is used to compute a pay raise: If a person worked 5 years or more, the pay raise is 10%. Otherwise, the pay raise is 8%.

Input: Current salary
              Years worked

Process: Step 1:   Are years worked $\geqslant$ 5?
                        Yes: Rate = .10
                        No: Rate = .08
              Step 2:   New salary = Current salary + Rate $\times$ Current salary
              Step 3:   Go to "Output"

Output: New salary

The flowchart for this algorithm would be written as follows:

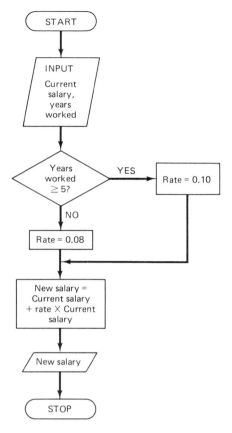

There can be more than one question in an algorithm. For example, an algorithm to compute the area and perimeter of a triangle must also include tests to identify a triangle. The test for a triangle is given as three sides where the sum of any two must be greater than the third. The algorithm would be:

Input: Sides of the triangle: A, B, C

Process: Step 1: Is $A + B > C$?
           Yes: Go to step 2
           No: Go to "output 2"
       Step 2: Is $B + C > A$?
           Yes: Go to step 3
           No: Go to "output 2"
       Step 3: Is $A + C > B$?
           Yes: Go to step 4
           No: Go to "output 2"
       Step 4: Perimeter $= A + B + C$
       Step 5: $S = $ perimeter $\div 2$
           Area $= \sqrt{S \times (S - A) \times (S - B) \times (S - C)}$
       Step 6: Go to "output, step 1"

Output 1:   Sides: A, B, C
                        Perimeter
                        Area
                        Stop

Output 2:   Not a triangle
                        Stop

Because the algorithm has two different possible outputs, a "stop" is specified at the end of each. This algorithm expressed as a flowchart would be

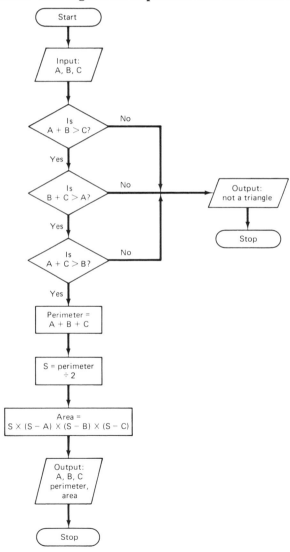

EXERCISES

Match statements 1 to 6 with the appropriate symbol.

A. ⬭   B. ▱   C. ▭   D. ◇

1. Interest = $P \times R \times T$
2. Output: Area
3. Start
4. Is time $\geqslant$ 5?
5. Time = Time + 1
6. Stop
7. Fill in the flowchart for the following algorithm, which changes time from a 24-hour clock to a 12-hour clock specifying A.M. or P.M.

   Input: Time
   Process: Step 1:   Is time = 12?
                          No: Go to step 2
                          Yes: Output: noon
             Step 2:   Is time < 12?
                          No: Go to step 3
                          Yes: Output: Time A.M.
             Step 3:   Time = Time – 12
             Step 4:   Output: Time P.M.

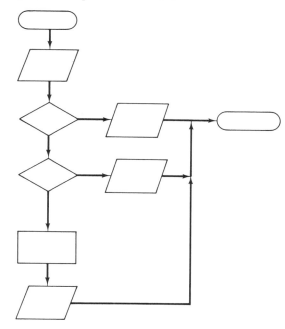

8. In Exercise 7 the algorithm involves a test for (= 12) and (< 12). Explain why step 3 does not test for (> 12).

9. Draw the flowchart for the following algorithm, which calculates the total salary for someone whose salary doubles daily for 30 days.

> **Input:**   Salary
>
> **Process:** Step 1:   Total salary = 0
>   Step 2:   Day = 1
>   Step 3:   Total salary = Total salary + Salary
>   Step 4:   Salary = 2 × Salary
>   Step 5:   Day = Day + 1
>   Step 6:   Is Day ≥ 30?
>   *Yes:* Go to "Output"
>   *No:* Go to step 3
>
> **Output:**   Total salary

10. Write the algorithm and draw a flowchart for computing the selling price given the following discounts.

Item	Discount
≤ $10.00	10%
More than $10.00 but less than $50.00	20%
Over $50.00	30%

FLOWCHARTING PROGRAMS WITH LOOPS

Loops are used to fill the tables, as well as for compounding or summations. Any type of problem that has repetitions can use a loop. Loops can be straight line or have decisions, or can even be inside other loops.

EXAMPLE 1

The markup on each item is 50%. The loop will fill in the retail price for each item without separate programming steps for each one.

Item no.	Wholesale cost	Retail price
243	$ 9.65	$_____
252	8.20	_____
288	2.45	_____
291	10.75	_____
342	7.99	_____
430	5.60	_____
540	1.98	_____
999 (test value for end of data)		_____

Solution   The algorithm for this problem would be as follows:

Input: Item number, wholesale cost

Process: Step 1:   Item number = 999?
                             Yes: Stop
                             No: Go to step 2
            Step 2:   Retail price = Wholesale cost + .50 × Wholesale cost
            Step 3:   Output—Retail price
            Step 4:   Go to "Input"

In this problem, for each input there is matching output, so the output statement is *in* the loop.

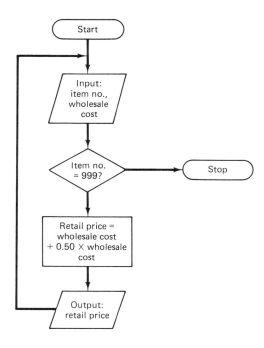

The loop will "execute" for any number of items. The last item number, which will identify the end of the list, will always be 999.

Another way to form a loop is with a counter. The counter has been defined as having these parts:

1. Initial value

2. Increment value

3. Test value

In flowcharting, a loop forms a circular pattern that stops because of an exit from the loop in the form of a decision statement. In flowcharting, a loop would look like this:

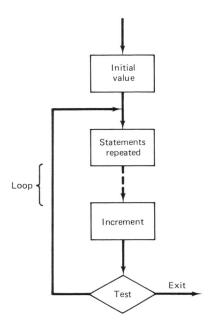

EXAMPLE 2

At a basketball game, tickets are 75 cents for children and $1.25 for adults. To aid the ticket seller, draw a table listing amounts for the following headings:

Number	Children	Adults
1	$ .75	$ 1.25
2	1.50	2.50
.	.	.
.	.	.
.	.	.
10	7.50	12.50

Solution   The algorithm is

Input: Amount for children = C
            Amount for adults = A

Process: Step 1:   Counter = 1
             Step 2:   Price for children = Counter $\times$ C
             Step 3:   Price for adults = Counter $\times$ A
             Step 4:   Output line of table
                       Counter, price for children, price for adults
             Step 5:   Counter = Counter + 1
             Step 6:   Is Counter > 10?
                          Yes: Stop
                          No: Go to step 2

The flowchart is

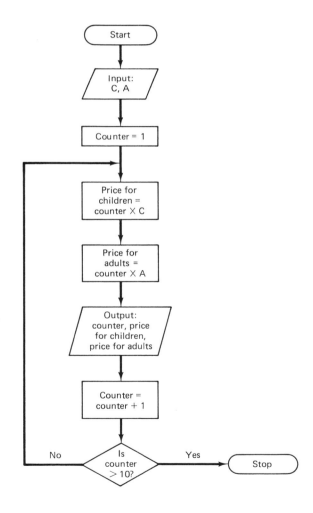

EXERCISES

Fill in the flowchart with the matching statements.

1.

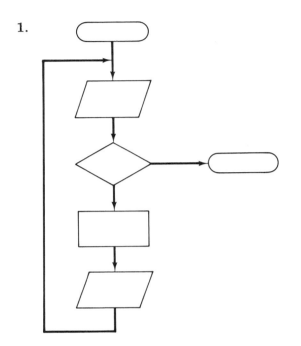

A. Output: Number,
    sales, commission
B. Commission =
    sales $\times$ .25
C. Stop
D. Input: Number,
    sales
E. Start
F. Is number = 999?

2. Draw a flowchart for the following algorithm.

Input: Employee number, hours worked, pay rate

Process: Step 1:   Is employee number = 999?
                           Yes: Stop
                           No: Go to step 2
              Step 2:   Are hours worked $>$ 40?
                           Yes: Gross pay = 40 $\times$ Pay rate + (hours worked
                                  – 40) $\times$ 1.5 $\times$ Pay rate
                           No: Gross pay = Hours worked $\times$ Pay rate
              Step 3:   Output: Line of table
                           Employee number, Gross pay
              Step 4:   Go to "Input"

3. Flowchart the following formula

$$PI = \frac{4}{1} - \frac{4}{3} + \frac{4}{5} - \frac{4}{7} + \frac{4}{9} - \frac{4}{11} + \frac{4}{13} - \frac{4}{15}$$

using this algorithm.

Process: Step 1: Sum = 0
Step 2: Numerator = 4
Step 3: Denominator = 1
Step 4: Fraction = Numerator ÷ Denominator
Step 5: Sum = Sum + Fraction
Step 6: Numerator = -1 × Numerator
Step 7: Denominator = Denominator + 2
Step 8: Is Denominator > 15?
          Yes: Go to "output"
          No: Go to step 4

Output: PI = Sum

4. Draw the flowchart for the following algorithm to calculate the value of *e*.

$$e = 1 + \frac{1}{1!} + \frac{1}{2!} + \frac{1}{3!} + \frac{1}{4!} + \frac{1}{5!} + \frac{1}{6!} + \frac{1}{7!} + \frac{1}{8!} + \frac{1}{9!} + \frac{1}{10!}$$

Process: Step 1: Sum = 1
Step 2: Factorial = 1
Step 3: Counter = 1
Step 4. Sum = Sum + 1/Factorial
Step 5: Counter = Counter + 1
Step 6: Factorial = Factorial × Counter
Step 7: Is Counter > 10?
          Yes: Go to "output"
          No: Go to step 4

Output: *e* = sum

5. Write the algorithm and draw the flowchart for calculating the sum

$$2 + 6 - 10 + 14 - 18 + 22 - 26 + 30 - 34 + 38$$

6. Write the algorithm and draw the flowchart to fill in the table using the formula

$$distance = .055 \times rate \times rate$$

Rate (miles per hour)	Stopping distance (feet)
10	_____
20	_____
30	_____
40	_____
50	_____
60	_____
70	_____
80	_____
90	_____
100	_____

SUMMARY

The algorithm is the first step, and flowcharting is the second step in developing a program. By following the boxes and working the algorithm in flowchart form, one can simulate computer operations. This is an important aid in program debugging before computer time is used.

CHAPTER REVIEW

1. The _____ is the pictorial representation of an algorithm.

Label each symbol.

2. ⬭ _____

3. ▭ _____

4. ▱ _____

5. ◇ _____

6. ↑ _____

Given the flowchart, answer Exercises 7 to 10.

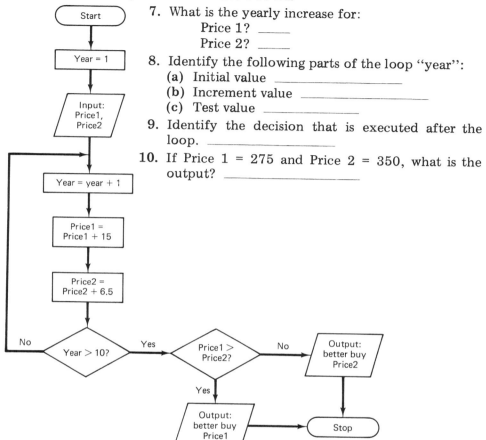

7. What is the yearly increase for:
   Price 1? _____
   Price 2? _____

8. Identify the following parts of the loop "year":
   (a) Initial value _____
   (b) Increment value _____
   (c) Test value _____

9. Identify the decision that is executed after the loop. _____

10. If Price 1 = 275 and Price 2 = 350, what is the output? _____

Language of Algebra

chapter 7

Algebra has certain characteristics that the programmer needs in order to set up the solutions for problems. In programming, the operations are on the right of the "=" sign and the results are stored in the variable on the left of the "=" sign.

The "=" sign can have two other meanings to the programmer. The first is "assign to" and is used to store the results of a calculation in a certain specified location (identified by a letter). The other meaning is "equal to" and is used in a relational expression. The relational expression is a comparison between two expressions and a test being made would be made for equality.

Because of these two meanings for the sign, algebraic expressions have certain differences from computer expressions. This chapter will stress the difference and use the computer expressions.

VARIABLES

Algebra is a number language. "Sentences" are written to solve problems which are similar to writing an algorithm. An algebraic sentence is written in general terms by using variables so that the sentence can apply to all problems of the same type.

A variable is a letter that represents some numeric quantity. The letter can have different values assigned according to the various problems used. For example, let I represent inches. If the problem is about the number of inches in a foot, then $I = 12$. If the problem is about the number of inches in a yard, then $I = 36$.

A constant is a number in an algebraic sentence that does not change. A sentence in algebra might be: $I = 12 \times F$. The variables are I for inches and F for feet. The constant is 12. The sentence means that the number of inches is 12 times the number of feet. (We use capital letters here because letters in computer expressions must always be capitals.)

OPERATIONS

The operations in algebra are:

> Exponentiation
> Multiplication
> Division
> Addition
> Subtraction

These operations are represented by certain symbols for algebra. But for the equivalent computer expressions, the symbols are different:

Algebra	Operation	Computer
+	Addition	+
–	Subtraction	–
$\times$	Multiplication	*
$\div$	Division	/
a^2	Exponentiation	↑ or **

In an algebraic expression, certain symbols for operations are understood. For example, *ab* means $a \times b$. *This is not true for computer expressions. All operations must be specified.*

HIERARCHY OF OPERATIONS

In any expression, there is a hierarchy to the order of calculations that *must* be followed. The order applies to both algebraic and computer calculations.

First: Exponentiation
Second: Multiplication or division from left to right
Third: Addition or subtraction from left to right

The order does make a difference—for example,

$$2 + 3 * 5 = ?$$

Following the order of calculations:

First: $3 * 5 = 15$

Second: $2 + 15 = 17$

The right answer is 17.
    Working the problem from left to right *ignoring* the order of calculations would give the wrong answer, as shown here:

First: $2 + 3 = 5$

Second: $5 * 5 = 25$

The wrong answer is 25.
    There is sometimes a need to change the order of calculations. To do this, add *parentheses* to the expression. The expression $2 + 3 * 5$ can be changed to $(2 + 3) * 5$. The order has now been changed so that the addition is done *first*, then the multiplication. Now, the correct answer is 25. Following are some examples.

(1) $6 + 15/3 + 4 = 6 + 5 + 4 = 15$

(2) $(6 + 15)/3 + 4 = 21/3 + 4 = 7 + 4 = 11$

(3) $(6 + 15)/(3 + 4) = 21/7 = 3$

In an algebraic format,

(1) $6 + \dfrac{15}{3} + 4$

(2) $\dfrac{6 + 15}{3} + 4$

(3) $\dfrac{6 + 15}{3 + 4}$

There are certain limitations imposed on computer expressions that can be ignored in algebraic expressions. The problems in this section will always use the following *computer limitations*.

1. All operation signs must be included.
2. Order of calculations is followed unless parentheses are included in an expression.
3. All computer expressions are one line.
4. Use only capital letters for variables.

Given the following algebraic expression:

$$\frac{cb + d}{a + b}$$

This expression must be rewritten as

$$(C * B + D)/(A + B)$$

EXERCISES

1. Identify the variables and constants in each expression.
   (a) $a + 4 + b$ (b) $cb + d$
   (c) $4 + 11 - e$ (d) $36f$
   (e) $(I/12) + 3Y$
2. Identify the operation in each expression.
   (a) Sum   (b) Difference
   (c) Quotient   (d) Product
   (e) Decrease   (f) Increase
   (g) Ratio   (h) Halve
   (i) Double   (j) Quarter
   (k) Plus
3. Write an algebraic sentence for each statement.
   (a) There are 36 inches in a yard.
   (b) The selling price is the cost increased by the profit.
   (c) The distance traveled is the product of the speed and time.
   (d) Density is the quotient of mass divided by volume.
   (e) The interest is rate times time, times principal.
   (f) The perimeter of a rectangle is double the sum of the length and width.
   (g) Overtime pay is 1.5 times the rate times the hours over 40.

4. Change each algebraic equation to one using one-line computer expressions and computer symbols.

   Example:
   Algebraic equation: $p = 2l + 2w$
   Computer expression and symbols: $P = 2 * L + 2 * W$

   (a) $a = 1 + prt$ **(b)** $\dfrac{q + p}{r - 4}$

   (c) $ax^2 + bx + c$ **(d)** $5xy + 4$

   (e) $\dfrac{a}{b} - \dfrac{c}{d}$ **(f)** $\dfrac{ab - c}{d}$

   (g) $\dfrac{ac}{bd}$ **(h)** $p(1 + rt)$

   (i) $b^2 - a^2$ **(j)** $(b - a)^2$

5. Identify the hierarchy of operations for each expression, and then give the answer for each.

   Example: $5 + 3 * 4/6 = 7$
              ③ ① ②

   (a) $9 + 5/2$ **(b)** $9 * 2/3 * 3$
   (c) $4 * 3 + 2$ **(d)** $13 + 2/5$
   (e) $8/2 * 9/3$ **(f)** $7 + 5 - 4 + 3$
   (g) $(14 - 8)/3 * 2$ **(h)** $4 * (3 * (2 + 1 * 4))$
   (i) $5 * (2 * (8 - 3) + 7)$ **(j)** $8/4 * 4/2$
   (k) $8/4/4/2$

ALGEBRAIC-TO-COMPUTER CONVERSION

Changing expressions involves knowing where to put parentheses and when.

1. $\dfrac{a + b}{c + d}$ In analyzing this algebraic problem, recognize that the numerator and denominator must be evaluated before the division takes place. Since the operations are addition, parentheses are necessary for a computer expression.

$$(A + B)/(C + D)$$

2. $\dfrac{a}{c} + \dfrac{b}{d}$ In a problem like this one, the division must be done *before* the addition. This is the order in which the problem will be evaluated, so parentheses are not necessary.

$$A/C + B/D$$

3. $\dfrac{ab}{c}$ The numerator is evaluated first, then division takes place. The problem does not need parentheses. However, the addition of parentheses in this case would not make the answer wrong. The problem becomes

$$A * B/C \qquad \text{or} \qquad (A * B)/C.$$

Both are correct.

4. $\dfrac{ab}{cd}$ Here, parentheses are necessary for a computer expression since with multiplication and division, evaluation is from left to right.

$$(A * B)/(C * D)$$

Another way of evaluating the problem is to see it as

$$\frac{a}{c} \times \frac{b}{d}$$

This becomes

(A/C) * (B/D)
① ③ ② ⟵ order of calculation

By removing the parentheses, the order of calculation is changed:

A/C * B/D
① ② ③ ⟵ new order of calculation

The order of calculation changes but the result is the same because

$$\frac{a}{c} \times \frac{b}{d}$$

can be

$$\frac{a}{c} \times b \times \frac{1}{d}$$

The same result can be obtained in the case of A * B/C/D, which is

$$ab \times \frac{1}{c} \times \frac{1}{d}$$

NESTED PARENTHESES

The set of characters used in a computer is limited, and parentheses are the only way to group these characters. By contrast, algebra has three ways to group its characters: parentheses ( ), brackets [   ], and braces {   }. This makes it easier to express an algebraic problem with multiple groupings.

An algebraic expression such as:

$$g - \{f[c(a+b) - d] + e\}$$

has to be rewritten as the following computer expression:

$$G - (F * (C * (A + B) - D) + E)$$

Whenever using multiple groupings or "nested parentheses," check for matching left and right parentheses.

EXERCISES

1. Rewrite as a computer expression.

(a) $3x + 5y$

(b) $5a + c - b$

(c) $a + b(d - 2)$

(d) $\dfrac{4a - 3b}{x + 2y}$

(e) $\dfrac{a}{5}c + 32$

(f) $\dfrac{5}{9}(f - 32)$

(g) $6 - \dfrac{x + 2}{5}$

(h) $\dfrac{x}{x + 2} - \dfrac{2}{x - 2}$

(i) $\dfrac{a + b + c}{2}$

(j) $b^2 - 4ac$

(k) $s(s - a)(s - b)(s - c)$

2. Write as computer expressions without using parentheses. Evaluate each using $a = 3$, $b = 4$, $c = 5$, and $d = 2$.

(a) $\dfrac{a + b}{c}$

(b) $c(a + b)$

(c) $\dfrac{ab}{c}$

(d) $\dfrac{abc}{bd}$

(e) $\dfrac{1}{2}abc$

POLISH NOTATION

Polish notation is a system of evaluating arithmetic expressions so that there is no need to worry about the order of operations or the use of parentheses. An expression in this system is written so that the operands precede their operators.

	Polish notation
A + B	A B +
C/D	C D /
4 * C	4 C *
5 – 11	(5) (11) –

Polish notation is a binary operation. An expression will always have two operands preceding the operator. With multiple operators such as A + B + C, the expression becomes A B C + +. The first operator (+) refers to the operands B and C. The second operator (+) refers to the sum of (B and C) and A. Always evaluate two at a time.

An expression with multiple operations such as A + B/C, according to the order of operations, would be evaluated:

1. B/C = X, then . . . .
2. A + X = the final result.

But in Polish notation, this becomes A B C / +.

The first operation is /, and the two operands immediately preceding this sign are done first: A B C / +. The result is then added to A, which is the second operation.

In an expression such as $(A + B)/C$:

1. $A + B = X$, then . . . .
2. $X/C = $ the final result.

In Polish notation, this becomes:

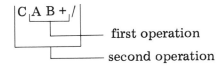

By writing this expression without parentheses, the order of evaluation is still correct.

To set up an expression with parentheses and multiple operations, apply the order of operations rule and set up the operations from left to right. For example:

$$A(B + C) - D$$

1.   Add B and C:

$$B \; C \; +$$

2.   Multiply by A:

$$A \; B \; C \; + \; *$$

3.   Subtract D:

$$D \; A \; B \; C \; + \; * \; -$$

It is not always possible to put all operands together. With certain expressions involving parentheses, it is necessary to mix operands and operators. For example:

$$(A + B)/(C * D)$$

1.   Add A and B:

$$A \; B \; +$$

2.   Multiply C and D:

$$C \; D \; *$$

3.   Divide the result of (1) by the result of (2) above:

$$A \ B + C \ D \ * \ /$$

Evaluate this expression for A = 10, B = 20, C = 3, and D = 2:

A B + C D * /	Add A and B	30
(30) C D * /	Multiply C and D	6
(30) (6) /	Divide 30 by 6	5

Calculations from left to right give the correct result.

When working a complex expression, the same procedure is followed. For example:

$$(A * B + C)/((A + B) * D)$$

1.   Multiply A and B:

$$A \ B \ *$$

2.   Add C to the result of (1) above:

$$C \ A \ B \ * \ +$$

3.   Add A and B:

$$A \ B \ +$$

4.   Multiply the result of (3) above by D:

$$D \ A \ B + \ *$$

5.   Divide the result of (2) above by the result of (4) above:

$$C \ A \ B \ * \ + \ D \ A \ B \ + \ * \ /$$

Evaluate for A = 5, B = 9, C = 7, and D = 2.

C A B * + D A B + * /	A multiplied by B	45
C(45) + D A B + * /	Add C and (45)	52
(52) D A B + * /	Add A and B	14
(52) D (14) * /	Multiply (14) by D	28
(52) (28) /	Divide (52) by (28)	1.857

Most computer languages involve translating from a high-level language to machine language. Polish notation would simplify the translation since parentheses and order of operations are not necessary.

EXERCISES

1. Translate into Polish notation.
   (a) C + D   (b) D * F + F
   (c) G – D + A   (d) (A + B + C) * D
   (e) (A + B * C)/(A * C)

2. Given A = 5, B = 4, C = 2, and D = 3, evaluate.
   (a) B C D + *   (b) B C * D +
   (c) A B C – /   (d) A C + A D – *
   (e) A B C A – * +   (f) A B * A (3) * (5) C * – +

3. Change to Polish notation and evaluate for X = 12, Y = 4, and W = 2.

$$\frac{XY - 5Y}{WX + Y}$$

4. Change to Polish notation.

$$(B – A) * (B + (A * (C – D) + B))$$

SUMMARY

The programmer uses certain properties common to algebra. But there are differences that are important for the programmer to understand. For example, algebraically, X = X + 7 makes no sense, but in computer notation, it means the number assigned to variable "X" is increased by 7 and reassigned to "X."

In algebra, variables are the letters that represent some numeric quantity. In computer notation, variables are the names of locations in computer memory where a number can be stored. Operations in both follow the same hierarchy. Parentheses are used to change the order of operations, or for clarity.

Polish notation is a method of writing expressions where the need for order of operations or parentheses is eliminated. This is a simplified method of writing calculations for computer evaluation.

CHAPTER REVIEW _____

1. The "=" sign has two meanings to the programmer. They are: _____ and _____ .

2. A _____ is a letter that represents some numeric quantity.

3. A _____ is a number in an algebraic sentence that does not change.

4. In the expression, F = 5280 X M, the variables are _____ , and the constant is _____ .

5. List the computer symbols for each operation.
   (a) Addition: _____
   (b) Subtraction: _____
   (c) Multiplication: _____
   (d) Division: _____
   (e) Exponentiation: _____

6. The hierarchy of operations specifies that calculations are in the following order:
   First: _____
   Second: _____
   Third: _____

7. _____ change the order of operations.

8. In the expression given here, write a number in the circle to show the order in which the operations would be executed:

$$A + (B * C + (D/2) - 5)$$
◯ ◯ ◯ ◯ ◯

9. The advantage of Polish notation is that _____ _____ .

10. The expression A B C D / + * is written algebraically as _____ .

Algebraic Expressions of "Not Equal"

chapter 8

In a computer program, there are two types of calculations: the first is "=" or *assignment* calculation, where the expression on the right is assigned to a variable on the left. The second type of calculation is the *conditional*. This is a check for basic number relationships. In this type of expression, no assignment is made; it is only a test for a certain condition.

In general, there are two types of conditions: "equal" or its reverse, "not equal." Under "not equal" are the following possibilities:

Less than, $<$
Less than or equal to, $\leqslant$
Greater than, $>$
Greater than or equal to, $\geqslant$

"Equal" expressions are called *equations*, and the "not equal" are called *inequalities*.

INEQUALITIES AND ALGEBRAIC SENTENCES

Inequalities can be expressed in other terms than just "greater than" or "less than."

If a minimum deposit is $10.00, algebraically it can be written: Deposit ≥ $10.00.

If a page can have at most 30 lines, it can be written: Page lines ≤ 30.

Inequality expressions can be:

Greater than	Less than
Minimum	Maximum
At least	At most
Exceeds	Cannot exceed

INEQUALITIES AND THE TRUTH SETS

An inequality is the expression of a "not equal" condition, and is written "≠." "Not equal" means one of the following:

>   Greater than

<   Less than

≥   Greater than or equal to

≤   Less than or equal to

The expression $X < 4$ has a truth set of integer values $\{\ldots 0, 1, 2, 3\}$. The real values cannot be specified since the set contains all the fractions between the whole numbers.

There are two methods of specifying truth sets of real numbers:

Method 1: Set notation

$\{X \mid X < 4\}$ would read: X "such that" X is less than 4.

Method 2: Graphing on a number line

The open circle means that 4 is *not included*, but all numbers less than 4, such as 3.9999, *are included.*

EXAMPLE 1

Given the expression $A \geqslant 5$, the truth set of integers is $\{5, 6, 7, 8, \ldots\}$. The truth set of real numbers is $\{A \mid A \geqslant 5\}$, or

This graph has the circle around 5 filled in, since the values are greater than *or equal to 5*. By filling in the circle, 5 *is included* in the truth set.

Some expressions must be simplified before they can be solved.

EXAMPLE 2

Solve $X + 4 < 5$.

Solution:

$$X + 4 < 5$$
$$\quad -4 \; -4 \quad \longleftarrow \text{subtract the same number from } both \; sides$$
$$X < 1$$

Truth set:

EXAMPLE 3

Solve $3X + 4 + X \leqslant 9$.

Solution:

$$3X + 4 + X \leqslant 9$$
$$4X + 4 \leqslant 9 \quad \longleftarrow \text{combine like terms}$$
$$\underline{\quad\quad -4 \; -4 \quad \longleftarrow \text{subtract the same number from both sides}}$$

$$4X \leqslant 5$$
$$\frac{4X}{4} \leqslant \frac{5}{4} \quad \longleftarrow \text{divide both sides by the same number}$$
$$X \leqslant 1\tfrac{1}{4}$$

Truth set:

NEGATIVES AND INEQUALITIES

On the number line, numbers to the left are the smallest. For example, consider two negatives such as -3 and -20. Since the -20 is to the left of -3, $-20 < -3$. The larger the absolute value of the negative, the smaller the number. Because of this relationship, two positive numbers can be reversed when they become negative. Consider the expression: $5 > 2$. If both are multiplied or divided by -1, the numbers become negative and the inequality is reversed, making the expression: $-5 < -2$.

In solving an inequality such as

$$\begin{array}{r} 10 - 4X > 34 \\ -10 \qquad\quad -10 \\ \hline -4X > 24 \end{array}$$

the next step is to divide both sides by a negative 4. This changes the inequality:

$$\frac{-4X}{-4} < \frac{24}{-4} = X < -6$$

On the number line, this would be

EXERCISES

1. Write as an algebraic sentence using symbols of inequality.
   (a) 8 is less than 15.
   (b) X can be at most 10.
   (c) The maximum number of deductions can be 15.
   (d) The age must be at least 16.
   (e) The deposit must exceed $1000.

2. Graph.
   (a) $D \leqslant 7$ (b) $X > 3$
   (c) $W \geqslant 5$ (d) $Y < 1.5$
   (e) $N > 2.8$

3. Specify the solution set for each graph.

   (a)

   (b)

   (c)

   (d)

4. Solve.
   (a) $X + 10 \leqslant -7$ (b) $Y - 6 \geqslant 2$
   (c) $2W + 4 > 14$ (d) $25 - 7X < 4$
   (e) $3X + 22 - 8X > 4$

TRUTH VALUES OF INEQUALITIES

Consider a statement such as $3X + 5 < X + 9$. To determine its truth value, given the set of X values, involves evaluating the expression and testing the inequality.

$$X \in \{0, 1, 2\}$$

$$3(0) + 5 \overset{?}{<} 0 + 9$$

$$5 \overset{?}{<} 9 \qquad \text{Yes}$$

$$3(1) + 5 \overset{?}{<} 1 + 9$$

$$8 \overset{?}{<} 10 \qquad \text{Yes}$$

$$3(2) + 5 \overset{?}{<} 2 + 9$$

$$11 \overset{?}{<} 11 \qquad \text{No}$$

The truth values are 0 and 1.
    Another way to solve the inequality is

$$
\begin{array}{rl}
3X + 5 &< X + 9 \\
-5 & \quad -5 \\
\hline
3X & < X + 4 \\
-X & \quad -X \\
\hline
\dfrac{2X}{2} & < \dfrac{4}{2} \\
X &< 2
\end{array}
$$

The numbers less than 2 are 1 and 0: the truth values for the inequality.

COMBINATIONS OF INEQUALITIES

Inequalities can be combined. A condition can specify that two inequalities must be satisfied to find a truth set. This is an *AND* condition. For example, X can be a minimum of 5 and a maximum of 8. The first condition is $X \geqslant 5$. The second condition is $X \leqslant 8$.

In programming, these conditions are combined with an "AND."

$$X \geqslant 5 \quad \text{AND} \quad X \leqslant 8$$

Graphically, these conditions are

The AND condition is true only if both inequalities are true, or where the two graph lines cross:

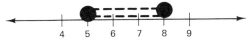

So, if 4 is less than 8 but not greater than 5, 4 is not in the solution set.

The AND condition can have no solution in certain situations, such as

$$X > 4 \quad \text{AND} \quad X < 2$$

No number can satisfy both conditions. Graphing shows that the arrow lines do not cross:

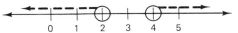

The reverse can happen with an *OR* condition. Given $A < 8$ OR $A > 5$. Since the graph is a *combination of both*, the solution set can be *any number* on the number line:

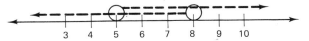

Another condition can be: X can be a maximum of 5 or a minimum of 8. The first condition is $X \leqslant 5$. The second condition is $X \geqslant 8$. But the combination is "OR":

$$X \leqslant 5 \quad \text{OR} \quad X \geqslant 8$$

This is graphically represented as the combination of both graph lines:

In this case, a number is not in the solution set if it is between 5 and 8.

The OR condition is true if either of the inequalities is true. The following table is used to compare the two types of conditions:

Condition 1	Condition 2	AND	OR
T	T	T	T
T	F	F	T
F	T	F	T
F	F	F	F

As the table indicates, the AND condition is only true if both conditions are true; the OR conditions is false only if both conditions are false.

OPPOSITES OF INEQUALITIES

A basic inequality such as $X > 10$ can also be written as $10 < X$. Here the first statement is reversed, but both are the same.

The opposite of $X > 10$ means that the other possible condition is $X \leqslant 10$. "12 is greater than 10" satisfies the original condition. It cannot satisfy the opposite condition, which is $X \leqslant 10$, and $12 \leqslant 10$ is false.

There are three basic relationships:

$$>$$
$$<$$
$$=$$

Therefore, the opposite of any one condition is one or both of the remaining ones.

Condition	Opposite
$>$	$\leqslant$
$\geqslant$	$<$
$<$	$\geqslant$
$\leqslant$	$>$

EXERCISES

1. Given the following inequalities, determine which of the specified values are in the truth set.

   (a) $4X - 6 < 3X + 8$ $\{7, 9, 25, 43\}$

   (b) $5X + 2 \geqslant 7X - 4$ $\{0, 1, 2, 3, 4, 5\}$

   (c) $6X - 3 + 2X \leqslant 12X - 8$ $\{-10, -5, 0, 5, 10\}$

   (d) $4X + 5 - 7X > 15 - 5X$ $\{2, 4, 6, 8, 10\}$

2. Graph the truth set for each combination.
    (a) $X \geqslant 5$ AND $X \leqslant 9$
    (b) $Y < 6$ AND $Y \geqslant 2$
    (c) $W > 10$ AND $W > 15$
    (d) $M < 4$ AND $M < 8$
    (e) $N > 7$ AND $N < 4$
3. Graph the truth set for each combination.
    (a) $A > 9$ OR $A < 5$
    (b) $N \leqslant 4$ OR $N \geqslant 2$
    (c) $M > 6$ OR $M > 8$
    (d) $Y < 0$ OR $Y < 1$
    (e) $C \geqslant 9$ OR $C < 1$
4. Give the opposite for each inequality.
    (a) $X > 4$ (b) $A \leqslant 7$
    (c) $B \geqslant 11$ (d) $W < 18$
    (e) $Y + 9 > 2Y$

SUMMARY _____

The opposite of an equality is an inequality. An inequality can be any one of the following:

> Less than: $<$
> Less than or equal to: $\leqslant$
> Greater than: $>$
> Greater than or equal to: $\geqslant$

Inequalities are used to test conditions, not to make assignments.

The solutions to an inequality, if they are integers, are simply the set of specified numbers. As real numbers, the numbers or fractions cannot be specified, so those solutions are usually graphed. The solutions of an inequality are known as its truth sets.

With an inequality, multiplying by a negative number will reverse the inequality. For example, if $9 > 7$, multiplying by a negative two (-2) gives $-18 < -14$.

Two inequalities can be combined with an AND or an OR statement. The AND condition is true if both inequalities are true, whereas the OR is false only if both inequalities are false.

An inequality is the opposite of an equality. Since there are multiple conditions of inequality, the opposite is always the remaining inequalities:

Inequality	Opposite
$>$	$\leqslant$
$\geqslant$	$<$
$<$	$\geqslant$
$\leqslant$	$>$

CHAPTER REVIEW

1. The opposite of an equality is _____ .
2. The inequality is basically a _____ expression.
3. The symbol for "less than" is _____ ; the symbol for "greater than" is _____ .
4. The word "minimum" is the same as the inequality _____ .
5. The word "maximum" is the same as the inequality _____ .
6. In graphing the solution set of an inequality, an open circle around a number means _____ , and a closed circle around a number means _____ .
7. If an inequality is multiplied or divided by a negative number, the inequality is _____ .
8. The AND condition is false if _____ .
9. The OR condition is true if _____ .
10. The opposite of $\geqslant$ is _____ .

Exponents

chapter 9

Understanding the use of exponents and their rules is important for a programmer, because exponents are used in many statements. Frequently, an algebraic statement must be interpreted into a statement the computer can understand. These statements must be constructed according to the constraints of the computer language, and this is where exponents are used.

EXPONENTIAL DEFINITIONS

An expression such as $A \times A \times A \times A \times A \times A$ can be shortened by using an exponent: A^6. In general, A is the base and 6 is the exponent. Looking at the problem just cited—$A \times A \times A \times A \times A \times A$—involves five calculations. A^6 involves one. The usual form for exponents in a computer language is A * * 6.

The inverse operation of exponentiation is finding the root. If 2 * * 5 = 32, the fifth root of 32 is 2. Algebraically, it is written as $\sqrt[5]{32} = 2$. The root of a number is also written as a fraction:

$$\sqrt[5]{32} = 32^{1/5}$$

Expression	Inverse	Root as a fractional power
$2^5 = 32$	$\sqrt[5]{32} = 2$	$32^{1/5} = 2$
$4^2 = 16$	$\sqrt{16} = 4$	$16^{1/2} = 4$

The second root is referred to as the *square root* and is symbolized by the radical sign ($\sqrt{}$) without any number.

Expression	Inverse	Root as a fractional power
$6^3 = 216$	$\sqrt[3]{216} = 6$	$216^{1/3} = 6$
$(3^2)^3 = 729$	$\sqrt{\sqrt[3]{729}} = 3$	$(729^{1/3})^{1/2} = 3$

Root expression	Root as a fractional power	Inverse
$\sqrt[5]{1024} = 4$	$(1024)^{1/5} = 4$	$4^5 = 1024$
$\sqrt{\sqrt[3]{46656}} = 6$	$((46656)^{1/3})^{1/2} = 6$	$(6^2)^3 = 46656$
$\sqrt[3]{8^5} = 32$	$(8^5)^{1/3} = 32$	$((32)^3)^{1/5} = 8$

Translated into a computer statement, our original example of $\sqrt[5]{32}$ is written as 32 * * (1/5).

In general:

$$\sqrt[N]{A} = A^{1/N} = A \ast\ast (1/N)$$

The power of $1/N$ is in parentheses because of the order of operations. Following this order, exponentiation precedes division. But with parentheses, division will be done first, as required by the order of operations.

A computer expression must be written as a single-line expression without the special root symbol. It is written using the fractional power, as illustrated below:

Algebraic	Computer
$\sqrt{25} = 25^{1/2}$	25 * * (1/2)
$\sqrt[4]{14641} = 14641^{1/4}$	14641 * * (1/4)
$\sqrt[5]{243} = 243^{1/5}$	243 * * (1/5)

An exponent that is expressed as a fraction can be simplified by changing it to its decimal equivalent: $1/2 = .5$; therefore, $9^{1/2} = 9^{.5}$. $1/4 = .25$; therefore, $16^{1/4} = 16^{.25}$. Those fractions that are repeating decimals are more accurate if left as fractions. For example, $1/3 = .33\overline{3}$ is a repeating nonterminating decimal. Depending on where the decimal is terminated, the result will change. To illustrate: $8^{1/3} = 2$, but

$$8^{.3} = 1.86607$$
$$8^{.33} = 1.98618$$
$$8^{.333} = 1.99861$$
$$8^{.3333} = 1.99986$$
$$8^{.33333} = 1.99999$$
$$8^{.333333} = 2$$
$$8^{.3333333} = 2$$

EXERCISES

1. Rewrite using exponents.
   (a) $7 \times 7 \times 7 \times 7 \times 7 \times 7 \times 7 \times 7$ (b) $CCCC$
   (c) $AABBBBBB$ (d) $(4)(4)(4)(4)AADDDD$

2. Find the inverse root operation.
   (a) $7^3 = 343$ (b) $2^8 = 256$
   (c) $12^2 = 144$ (d) $(5^3)^2 = 15625$
   (e) $\sqrt[4]{2401} = 7$ (f) $\sqrt[3]{1728} = 12$
   (g) $\sqrt{\sqrt[4]{100000000}} = 10$ (h) $\sqrt[5]{\sqrt[3]{\sqrt{1}}} = 1$
   (i) $\sqrt{16^3} = 64$ (j) $\sqrt[4]{5^8} = 25$
   (k) $\sqrt[3]{\sqrt[3]{262144}} = 4$ (l) $\sqrt{\sqrt[5]{3^{20}}} = 9$

3. Complete the table.

Root expression	Root as a fractional power	Root as a decimal power
(a) $\sqrt[5]{725}$	_____	_____
(b) $\sqrt{1600}$	_____	_____
(c) _____	$(50625)^{1/8}$	_____
(d) _____	$(2197)^{1/3}$	_____
(e) _____	_____	$(412)^{.125}$
(f) _____	_____	$(279841)^{.25}$
(g) _____	_____	$(324)^{.5}$
(h) $\sqrt[5]{\sqrt{1024}}$	_____	_____
(i) _____	$((625)^{1/2})^{1/4}$	_____
(j) _____	_____	$((6561)^{.5})^{.25}$

4. Rewrite as a computer expression.
   - (a) $\sqrt[3]{27}$ (b) $\sqrt{625}$
   - (c) $\sqrt[4]{81}$ (d) $\sqrt[6]{4096}$
   - (e) $\sqrt[7]{1}$

5. Using a calculator, evaluate each expression. (*Note:* $3^6 = 729$.)
   - (a) $729^{.16} =$ (b) $729^{.166} =$
   - (c) $729^{.1666} =$ (d) $729^{.16666} =$
   - (e) $729^{.166666} =$ (f) $729^{.1666666} =$
   - (g) $729^{.17} =$ (h) $729^{.167} =$
   - (i) $729^{.1667} =$ (j) $729^{.16667} =$
   - (k) $729^{.166667} =$ (l) $729^{.1666667} =$

RULES OF EXPONENTS

The rules of exponents are used to simplify expressions. The rules are:

1. $(A^M)(A^N) = A^{M+N}$

2. $(A^M)^N = A^{MN}$

3. $\dfrac{A^M}{A^N} = A^{M-N}$

4. $\dfrac{1}{A^M} = A^{-M}$ $(A \neq 0)$

5. $(AB)^M = A^M B^M$

6. $\left(\dfrac{A}{B}\right)^M = \dfrac{A^M}{B^M}$ $(B \neq 0)$

7. $\sqrt[N]{A^M} = A^{M/N}$

8. $\sqrt[N]{A}\,\sqrt[N]{B} = \sqrt[N]{AB}$

9. $\dfrac{\sqrt[N]{A}}{\sqrt[N]{B}} = \sqrt[N]{\left(\dfrac{A}{B}\right)}$

These rules can be used to simplify a statement or to decrease the number of operations. For example: $\sqrt{5^4}$ using rule 7 becomes $5^{4/2}$, reducing the fraction $4/2 = 2$. The simplified problem:

$$\sqrt{5^4} = 5^{4/2} = 5^2$$

In certain cases, the problem can be very complicated if translated from the algebraic to a computer expression. The rules (above) can be used to make the problem clearer. For example,

$$\sqrt{8^2 A^4 B^2} = (8^2 A^4 B^2)^{1/2}$$

as a computer expression becomes

$$(8**2 * A**4 * B**2)**(1/2)$$

This requires seven operations. Simplify this by first using rules 7 and 5:

$$(8^2 A^4 B^2)^{1/2} = 8^{2/2} A^{4/2} B^{2/2} = 8A^2 B$$

which is simply

$$8 * A**2 * B$$

requiring only three operations.

PROBLEM SOLVING WITH EXPONENTS

Programmers writing programs for the various economic and scientific disciplines frequently need to translate a formula into a computer expression. For example, a standard compound interest formula is

$$A = \text{principal and interest}$$
$$N = \text{number of years}$$
$$I = \text{rate of interest}$$
$$P = \text{principal}$$
$$A = P(1 + I)^N$$

As a computer expression, it becomes

$$A = M * ((1 + R)**(1/M) - 1)$$

Another standard formula for annual rate when interest is compounded more than once a year is

$$M = \text{number of times a year interest is compounded}$$
$$R = \text{rate at which interest is compounded}$$
$$A = \text{annual rate}$$
$$A = M(\sqrt[M]{1 + R} - 1)$$

As a computer expression, it becomes

$$A = M * ((1 + R) * * (1/M) - 1)$$

And another standard formula for finding a rate for a calculation of more than 1 year:

$$P = \text{principal}$$

$$F = \text{accumulated value}$$

$$Y = \text{number of years}$$

$$R = \text{interest rate}$$

$$R = \sqrt[Y]{\frac{F}{P}} - 1$$

As a computer expression, it becomes

$$R = (F/P) * * (1/Y) - 1$$

EXERCISES

1. Using the rules of exponents, simplify each expression.
   (a) $(P^5)(P^2)(P^3)$ (b) $(R^4)^3$
   (c) S^{12}/S^9 (d) T^5/T^{16}
   (e) $(A^5/D^6)^2$ (f) $(X^4 Y^3 W^5)^2$
   (g) $\sqrt[3]{A^5 B^6} / \sqrt[3]{A^2 B^{12}}$ (h) $\sqrt[4]{X} \sqrt[4]{X^2} \sqrt[4]{X^5}$
   (i) $\sqrt{27C^3 D^5} \sqrt{3CD^3}$ (j) $\sqrt[3]{15AB^4 D^2} \sqrt[3]{5^2 AC^3 D^2} \sqrt[3]{3^5 AB^2 D^2}$

2. Rewrite as a computer expression.
   (a) $5^2 A^3 B$ (b) $\sqrt{225X^2 Y}$
   (c) $\sqrt[3]{27YW}$ (d) $\sqrt{RW^3 Z}$
   (e) $\sqrt[5]{X^7}$

3. Change to a computer expression.

$$\frac{\sqrt{27X^3 Y^5}}{\sqrt{3XY^2}}$$

   (a) Count the number of calculations required.
   (b) Simplify the expression.
   (c) Count the number of calculations in the simplified expression.

4. Rewrite as a computer expression.

(a) $D = \dfrac{R\sqrt{N-2}}{\sqrt{1-R^2}}$ (b) $V = {}^{M+N}\!\!\sqrt{\dfrac{BM}{AN}}$

(c) $T = 2\pi \sqrt{\dfrac{L}{32.2}}$ (d) $K = \sqrt{S(S-A)(S-B)(S-C)}$

(e) $V = \sqrt{\dfrac{L+2M}{P}}$ (f) $G = K\sqrt{L^2 + (X-R)^2} - \sqrt{L^2 + (X+R)^2}$

SUMMARY

Computer expressions use * * to raise to a power. Roots must be changed to an exponent since the radical symbol is usually not available on a keyboard. Therefore, expressions with roots are converted to fractional powers first and then written using the power symbols.

Radicals are frequently complicated expressions which can be simplified, thus allowing for fewer calculations.

CHAPTER REVIEW

1. In the expression C^6, C is _____ and 6 is _____.
2. As a computer expression, C^6 is written as _____.
3. The symbol $\sqrt{}$ is called a _____.
4. The root of a number is a _____ power.
5. The inverse operation of raising to a power is finding the _____.
6. The inverse operation of taking the root of a number is _____ _____.
7. The square root of a number is the _____ power of the number.
8. A fraction can be changed into a _____ by dividing denominator into numerator.
9. A repeating nonterminating decimal is _____ if left as a fraction.
10. The rules of exponents are used to _____ expressions.
11. An expression such as A * * (1/4) means _____ root of _____.
12. In A * * (1/4), the base is _____ and the power is _____.

Equations

A computer programmer deals with equality on two levels. On one, the equality is really an *assignment*. If a programmer writes the assignment sentence as "A = B + 3," it means that the value of B, increased by 3, is assigned to A. Because of this, a statement like "A = A + 1" just means to increase the value of A by 1 and store that new value in location A.

The other level is the *conditional*. Here the meaning of A = B + 3 is the comparison between the value of A and the value of B + 3. If they are the same, the conditional is true. Otherwise, the conditional is false.

In an assignment statement, the format is: single variable on the left, algebraic expression on the right. The calculated value of the expression is assigned to the variable. Much of equation solving is manipulation of terms so that the variable on the left is the one for which a solution is to be found.

OPERATIONS WITH SIMPLE EQUATIONS

An algebraic sentence is an equation. The solution to an equation is determined by the type. There are basically four types of simple equations:

    1.   X + A = B
    2.   X − A = B
    3.   X/A = B
    4.   X * A = B

In each case the solution involves finding the value of X, the value on the left. To solve for X, use the inverse operation:

Operation	Inverse
+	−
−	+
/	*
*	/

To solve type 1, X + A = B, note that the operation here is "+," the inverse of "−." Therefore, subtract A from both sides of the "=" sign:

$$X + A = B$$
$$\underline{\quad - A \quad - A \quad}$$

Result: $X = B - A$

Since: $A - A = 0$ and $X + 0 = X$

To solve type 2, again use the inverse and add A to both sides of the "=" sign:

$$X - A = B$$
$$\underline{\quad + A \quad + A \quad}$$

Result: $X = B + A$

Since: $- A + A = 0$ and $X + 0 = X$

To solve type 3, use the inverse and multiply both sides by A

$$X/A = B$$
$$\underline{\quad * A \quad * A \quad}$$

Result: $X = B * A$

Since: $\dfrac{1}{A} * A = 1$ and $X/1 = X$

To solve type 4, use the inverse and divide both sides by A:

$$X * A = B$$
$$/ A \quad / A$$

Result: $X = B/A$

Since: $A/A = 1$ and $X * 1 = X$

Most equations of type 4 are written in the form $XA = B$, but they can be changed to the computer notation of $X * A = B$.

EQUATIONS WITH COMBINATIONS

Most equations are not simple equations. Instead, they are a combination of all four types. To solve these equations, *reverse the operations.*

EXAMPLE 1

$$X/A + B = C$$

Solution   Flowcharting this equation as it stands is

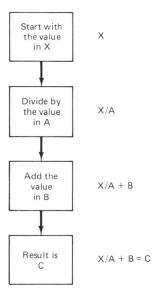

To find the solution, start with the result and *reverse the steps, taking the inverse of each operation.*

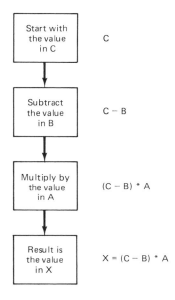

Using the inverse method directly (without flowcharting), start with the inverse of the last operation. Normally, in the sample equation, the order of operations dictates that division is first and addition is last. But with the inverse method, begin with the inverse of the last operation:

$$\begin{array}{c}
X/A + B = C \\
\underline{\quad - B \quad - B \quad} \\
X/A = C - B \\
\underline{\quad * A \quad * A \quad} \\
X = (C - B) * A
\end{array}$$

Either the flowcharting or the direct method can be applied.

EXAMPLE 2

Given the equation $5X - 3 = 17$. This is rewritten as a computer expression as $5 * X - 3 = 17$.

Solution using the
flowcharting method

Solution using the
inverse method

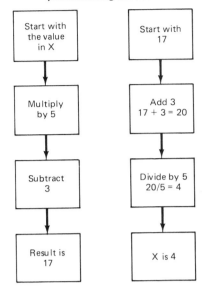

$$X * 5 - 3 = 17$$
$$+ 3 \quad + 3$$

$$X * 5 = 20$$
$$/ 5 \quad / 5$$

$$X = 4$$

EXERCISES

1. Solve.
   (a) $X + C = D$ (b) $X - 5 = A$ (c) $X/Y = 8$
   (d) $X * 7 = F$ (e) $4X = G$ (f) $X + 17 = Y$
   (g) $X - D = 11$ (h) $X/5 = P$ (i) $9X = AB$
   (j) $X * A = C$

2. Find the value of X.
   (a) $X + 9 = 71$ (b) $X - 82 = 112$ (c) $X/12 = 3024$
   (d) $14X = 98$ (e) $X * 5 = 1070$

3. Solve using the flowchart or inverse method.
   (a) $3X - 11 = 157$ (b) $X/7 - 8 = 10$ (c) $15X + 75 = 135$
   (d) $X/23 + 21 = 26$ (e) $3X/8 + 11 = 17$ (f) $5X/12 - 4 = 6$

4. Show the solution to $CX/D + A = G$:
   (a) Using the flowchart method.
   (b) Using the inverse method.

SOLVING FORMULAS FOR SPECIFIC VALUES

Methods of equation solving can be used to establish formulas for certain solutions.

EXAMPLE 1

The formula for interest earned is $I = PRT$, where

$$P = \text{amount}$$
$$R = \text{rate}$$
$$T = \text{time in years}$$

Solution   A problem to find time, given interest, rate, and amount, would involve changing the formula with T on the left.

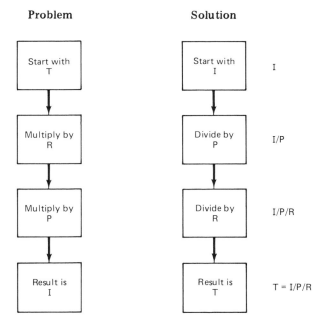

The formula is now $T = I/P/R$. This can be done with any given formula.

Those formulas with multiple operations can be solved by using the same method.

EXAMPLE 2

The perimeter of a rectangle is given as $P = 2L + 2W$, where

$$L = \text{length}$$
$$W = \text{width}$$
$$P = \text{perimeter}$$

Solution   Solving for W:

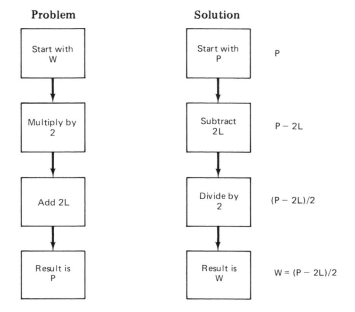

EXAMPLE 3

A formula such as the Pythagorean theorem $C^2 = A^2 + B^2$ since it involves squaring, will have roots in the solution.

Solution   Solving for B:

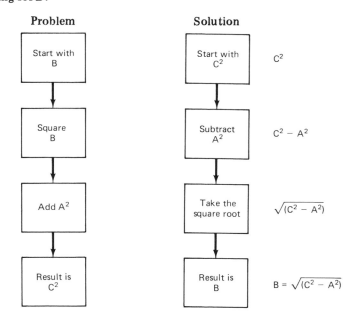

The solution $B = \sqrt{(C^2 - A^2)}$ algebraically can be translated into a computer expression

$$B = (C ** 2 - A ** 2) ** (1/2)$$

EXAMPLE 4

The formula for Centigrade temperature, given Fahrenheit, is $C = \frac{5}{9}(F - 32)$.

Solution   This easily becomes the formula for F, given C.

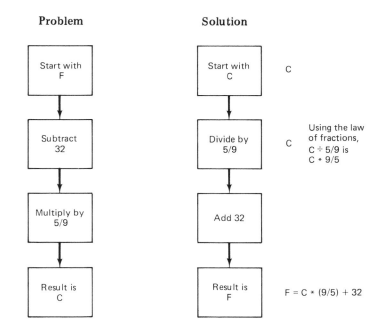

EXERCISES

1. Solve each formula for the specified variable.
   (a) $A = LW$ L
   (b) $A = \frac{1}{2}BH$ H
   (c) $V = \pi R^2 H$ R
   (d) $A = \frac{1}{2}H(B + C)$ C
   (e) $A = \dfrac{(P - L) + K}{N}$ P
   (f) $I = \sqrt[N]{F/P} - 1$ F
   (g) $M^2 = KT$ K
   (h) $A^2 + B^2 = C^2$ A
   (i) $V = 125.3\sqrt[6]{ZT}$ T
   (j) $T = 2\pi\sqrt{L/32.2}$ L

RATIO EQUATIONS

One type of equation is the *ratio equation*. The ratio equation has one unknown and three givens. Using this information, one can solve for the missing number.

EXAMPLE 1

If computer time charges are $35 per hour, how much will 16 minutes cost?

Solution:

$$\text{Unknown cost} = X \text{ per 16 minutes}$$

$35 per hour can be written as the ratio $\frac{35}{60}$ minutes

$$X \text{ per 16 minutes} = \frac{X}{16}$$

Both ratios must be the same when reduced.

$$\frac{35}{60} = \frac{X}{16}$$

1. Find the cross-product.

$$\frac{35}{60} \diagup\!\!\!\!\diagdown \frac{X}{16}$$

$$60X = 35(16)$$
$$= 560$$

2. Divide both sides by 60.

$$\frac{60X}{60} = \frac{560}{60}$$

$$X = \$9.33$$

3. Cost of 16 minutes of computer time is $9.33.

The basic ratio equation is

$$\frac{X}{A} = \frac{B}{C}$$

The solution is

$$CX = AB \qquad X = \frac{AB}{C}$$

The rules are:

1. Both ratios have identical units.
2. Both ratios must be set up consistently.

EXAMPLE 2

If it takes 4 minutes to print paychecks for 100 employees, how long will it take to print paychecks for 800 employees?

Solution:

$$\text{Using } \frac{\text{minutes}}{\text{number of employees}} \quad \frac{4}{100} = \frac{X}{800} \qquad 3200 = 100X, \qquad 32 \text{ minutes} = X$$

$$\text{Using } \frac{\text{number of employees}}{\text{minutes}} \quad \frac{100}{4} = \frac{800}{X} \qquad 3200 = 100X, \qquad 32 \text{ minutes} = X$$

In both approaches, consistency of numerator and denominator is maintained. In the first approach, all numerators are "minutes," and all denominators are "number of employees." Similarly, in the second approach, all numerators are "number of employees," and all denominators are "minutes."

Problem solving with ratios is directly applicable to percentage problems.

EXAMPLE 3

If on a test, 85% meant that 15 questions were correctly answered, how many questions were on the test?

Solution   Percentage is a ratio over 100, so 85% is $\frac{85}{100}$. Therefore,

$$\frac{85}{100} = \frac{15}{X} \qquad \text{or} \qquad \frac{\text{number right}}{\text{Total}}$$

$$85X = 1500$$

$$X = 17.647 \quad \text{or} \quad 18 \text{ questions on the test}$$

The same approach can be used if the problem were changed.

EXAMPLE 4

If on a test of 15 questions, 85% were right, how many were right?

Solution   In this problem, the total is known, but not the number correct.

$$\frac{85}{100} = \frac{X}{15}$$

$$100X = 1275$$

$$X = 12.75 \quad \text{or} \quad 13 \text{ correct}$$

EXERCISES

1. If it takes 8 hours to process the reports manually for four drivers, how long will it take for six drivers?

2. Print time is 300 lines per minute. If payroll takes 2500 lines, how long will it take to print the payroll?

3. A keypunch operator can type 500 cards per hour. How long will it take to type 3200 cards?

4. Equipment is depreciated by $\frac{1}{6}$ of its value at the end of the first year. If the purchase price was $4500, what was its value at the end of the first year?

5. A disk rotates at a rate of 2400 revolutions per minute. How many revolutions does it make in a half-hour?

6. Expenses for office supplies ran $250 per month for 30 employees. If the staff is decreased by 8 employees, how much can be saved on supplies?

7. The cost of office space for 560 square feet is $12,320. If a company needs 120 more square feet, what will be the new cost?

8. A sort routine for 100 records runs .9 second. Another sort routine runs 2.6 seconds for 500 records. Which would run the most efficiently for 25,000 records?

9. One model of a magnetic tape runs at a tape speed of 125 inches per second, with a density (bytes per inch) of 800 bytes. A second model runs a tape at the speed of 200 inches per second, with a density of 1600 bytes per inch. Of the two, which has the faster data transfer rate (bytes per second)?

10. A $5\frac{1}{4}$-inch floppy disk has a data transfer rate of 12,500 bytes per second, 35 tracks, 10 sectors per track, and 256 bytes per sector. What are the number of bytes
    (a) Per track?
    (b) Per diskette?

11. A company has available $4.32 for every $1 of current liability. If the total assets of the company are $10,428,600, what are the total liabilities?

12. An efficiency ratio is the number of forms processed over the standard number of forms. If a worker has an efficiency ratio of 130 and has processed 750 forms, what is the standard?

13. In a file of 1200 records, each record is 225 bytes. Find:
    (a) Records per track if stored on a disk with 13,000 bytes per track.
    (b) Number of tracks needed.

14. Given the following data fields, identify the ratios that can be computed for each.

(a) Average number of employees; weekly cost

Ratio = _____

(b) Total number of employees; total lines of print

Ratio = _____

(c) Total number of lines printed; number of minutes for run

Ratio = _____

(d) Total cost of training; total number trained

Ratio = _____

SUMMARY

Solving equations can be done by the computer in most cases. The only problem facing the programmer is to establish the equation for solution. This means writing the equation to be solved on the right, and variables for solution on the left of the "=" sign.

Equation manipulation is simply reversing the steps shown in the original equation. Flowcharting can be helpful. Formulas are just equations with variables. The formula can be manipulated in order to solve for any variable.

Solving an equation is a matter of finding some unknown quantity, given certain values. Ratio equations are one way to solve for unknowns using fraction ratios.

CHAPTER REVIEW

1. Given, C = A * B means the value of _____ is to be assigned the value of _____ .
2. Finding X in the equation $B = X + A$ involves _____ A.
3. To solve an equation, use the _____ of the operation specified.
4. Solve X * 3 + B = D using the flowchart method.

Problem   Solution

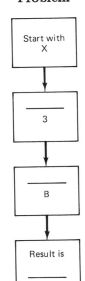

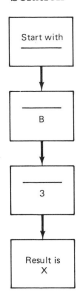

5. Solving $X + 11 = 17$ using the inverse method involves _____ _____ from both sides of the equation.

6. If $I = PRT$, the formula for R is _____ .

7. Given the ratio

$$\frac{X}{A} = \frac{B}{C}$$

the solution is $X =$ _____ .

8. If one ratio is in miles/hour, the other ratio must be in _____ .

Quadratic Equations

Equations of the form $Ax^2 + Bx + C = 0$ are called *quadratic equations*. The variable is x and the highest power of x is **2**. Because of the x^2, these equations can have at most two values for x as solutions or roots. This chapter focuses on understanding these equations.

STANDARD FORM

Any equation with a single squared term is a quadratic, but not all quadratic equations are in standard form. The *standard form* of the quadratic is

$$Ax^2 + Bx + C = 0$$

Any quadratic equation can be changed to standard form.
    Given the quadratic

$$5x^2 = 7x + 3$$

Changing it to standard form becomes

$$5x^2 - 7x - 3 = 0, \qquad \text{where } A = 5, B = -7, C = -3$$

Given the quadratic

$$4x^2 + 10 = 8x - 3$$

Changing it to standard form becomes

$$4x^2 - 8x + 13 = 0, \qquad \text{where } A = 4, B = -8, C = 13$$

ROOTS OF THE QUADRATIC

The general form of the quadratic is $Ax^2 + Bx + C = 0$. When this equation equals zero, it can be solved for x. The values for x are the roots of the equation. The quadratic is generally formed by multiplication of two linear equations. In that form, the roots can be found by observation.

For example, the multiplication problem $(x + 4)(x + 2) = 0$, when multiplied, becomes

$$x^2 + 2x + 4x + 8 = 0$$

or in standard form,

$$x^2 + 6x + 8 = 0.$$

To find the roots, use the *rule of zero:*

$$AB = 0 \qquad \text{if} \qquad A = 0 \quad \text{or} \quad B = 0$$

In the case of the problem cited above,

$$(x + 4)(x + 2) = 0 \qquad \text{if} \qquad x + 4 = 0 \quad \text{or} \quad x + 2 = 0$$

which means that the roots can be $x = -4$ since $-4 + 4 = 0$, or $x = -2$ since $-2 + 2 = 0$.

To verify the quadratic equation for $x = -4$:

$$(-4)^2 + 6(-4) + 8 = 16 - 24 + 8 = 24 - 24 = 0$$

To verify the quadratic equation for $x = -2$:

$$(-2)^2 + 6(-2) + 8 = 4 - 12 + 8 = 12 - 12 = 0$$

Quadratics can have one root when they are of the form $(x + 3)^2 = 0$. In this case the root is -3.

A further illustration: in the multiplication problem, $(3x - 2)(4x - 1) = 0$, the roots must be calculated:

$$3x - 2 = 0 \quad \text{or} \quad 4x - 1 = 0$$
$$3x = 2 \quad \text{or} \quad 4x = 1$$
$$x = \tfrac{2}{3} \quad \text{or} \quad x = \tfrac{1}{4}$$

EXERCISES

1. Identify which of the following are quadratic equations.
   (a) $7x + 11 = 15 + 3x$
   (b) $4x^2 - 7x + 15 = 0$
   (c) $15x^2 + 3x = 7$
   (d) $9x^2 = 17x + 23$
   (e) $5(x + y) = 7x$
   (f) $8x(x + 2) = 24$
   (g) $5x(2x - 3) = 10x^2 + 12$
   (h) $7(4x^2 - 3x + 4) = x$
   (i) $2x(4x + 3) = x(9 - 8x) + 21$
   (j) $3(5x^2 + 7x - 15) = x(4x^2 - 3x + 11)$

2. Change into standard form.
   (a) $8 + 3x = 5x^2$
   (b) $7x^2 + 9 = 4x^2 + 3x$
   (c) $12x^2 - 3x + 8 = 25$
   (d) $12x^2 + 4x - 9 = 3x^2 + 5$
   (e) $9x^2 + 7x + 2 = 18$
   (f) $4(x + 2) = x(3x + 7)$
   (g) $11x^2 - 9x = 27x - 29$
   (h) $5x(x - 7) = 22$
   (i) $4x^2 - 7x(8 - 2x) = 6(x + 3)$
   (j) $15 - 12x^2 + 5x(x - 4) = 29$

3. Find the roots.
   (a) $(x + 7)(x + 5) = 0$
   (b) $(x + 9)(x + 8) = 0$
   (c) $(x + 8)(x - 3) = 0$
   (d) $(x - 71)(x + 2) = 0$
   (e) $(x - 12)(x - 11) = 0$
   (f) $(x + 9)^2 = 0$
   (g) $(x - 14)^2 = 0$
   (h) $(3x + 5)(x - 12) = 0$
   (i) $(2x - 9)(5x - 17) = 0$
   (j) $(7x + 3)(8x + 5) = 0$

SUM AND PRODUCT OF ROOTS

In general, the roots of a quadratic equation can be identified as r and s. With this identification, the problem is expressed as $(x - r)(x - s) = 0$. Multiplying gives

$$(x - r)(x - s) = x^2 - sx - rx + rs = 0$$
$$= x^2 - (s + r)x + rs = 0$$

If the coefficient of x^2 is 1, the sum of the roots is the coefficient of x or $(s + r)$, and the product of the roots is the constant rs.

Using the standard form of the quadratic equation

$$Ax^2 + Bx + C = 0$$

the sum and product of the roots can be found by changing the equation as illustrated:

Dividing each term by A,

$$\frac{A}{A} x^2 + \frac{B}{A} x + \frac{C}{A} = \frac{0}{A}$$

results in

$$x^2 + \frac{B}{A} x + \frac{C}{A} = 0$$

Therefore, the sum of the roots,

$$-(r + s) = \frac{B}{A} \qquad \text{or} \qquad r + s = -\frac{B}{A}$$

The product of the roots,

$$rs = \frac{C}{A}$$

EXAMPLE 1

Given the equation $x^2 - 7x + 12 = 0$. The sum of the roots is -7; the product of the roots is 12. Find the roots.

Solution   Find the possible combination of factors that have 12 as a product and 7 as a sum:

r	s	$r \times s = 12$?	$r + s = 7$?
1	12	Yes	No
2	6	Yes	No
3	4	Yes	Yes

The roots are 3 and 4.

EXAMPLE 2

Find the sum and product of the roots of $7x^2 - 24x + 20 = 0$.

Solution   The sum of the roots is

$$r + s = -\frac{B}{A}$$

$$= -\left(-\frac{24}{7}\right)$$

$$= \frac{24}{7}$$

The product of the roots is

$$rs = \frac{C}{A}$$

$$= \frac{20}{7}$$

EXAMPLE 3

Given the roots $r = \frac{2}{5}$ and $s = \frac{4}{5}$, find an equation of the form $Ax^2 + Bx + C = 0$.

Solution:

$$r + s = \frac{2}{5} + \frac{4}{5} = \frac{6}{5}$$

$$rs = \left(\frac{2}{5}\right)\left(\frac{4}{5}\right) = \frac{8}{25}$$

Substituting in the equation gives

$$x^2 - (s + r)x + rs = 0$$

$$x^2 - \frac{6}{5}x + \frac{8}{25} = 0$$

Multiplying each side by 25:

$$25(x^2) - 25\left(\frac{6}{5}\right)x + 25\left(\frac{8}{25}\right) = 25(0)$$

$$25x^2 - 30x + 8 = 0$$

EXAMPLE 4

Given the roots $r = -\frac{1}{3}$ and $s = \frac{2}{7}$, find the equation of the form $Ax^2 + Bx + C = 0$.

Solution:

$$r + s = -\frac{1}{3} + \frac{2}{7} = \frac{-7 + 6}{21} = -\frac{1}{21}$$

$$rs = \left(-\frac{1}{3}\right)\left(\frac{2}{7}\right) = -\frac{2}{21}$$

Substituting in the equation gives

$$x^2 - (s + r)x + rs = 0$$

$$x^2 - \left(-\frac{1}{21}\right)x + \left(-\frac{2}{21}\right) = 0$$

$$x^2 + \frac{1}{21x} - \frac{2}{21} = 0$$

Multiply each side by 21:

$$21x^2 + x - 2 = 0$$

EXERCISES

1. Change into standard form using the sum of the roots and the product of the roots.
   (a) $(x - 8)(x - 3) = 0$ (b) $(x + 5)(x + 12) = 0$
   (c) $(x + 12((x - 7) = 0$ (d) $(x - 3)(x + 1) = 0$
   (e) $(x + 27)(x - 27) = 0$

2. Find the sum and product of the roots.
   (a) $x^2 - 18x + 77 = 0$ (b) $x^2 - 7x + 12 = 0$
   (c) $x^2 + 11x + 30 = 0$ (d) $4x^2 + 12x + 24 = 0$
   (e) $5x^2 + 22x + 21 = 0$ (f) $6x^2 - 21x + 35 = 0$
   (g) $21x^2 - 17x + 2 = 0$ (h) $6x^2 - 3x - 3 = 0$
   (i) $15x^2 - 14x - 16 = 0$ (j) $21x^2 + 13x - 18 = 0$

3. Using sum and product formulas, find the roots.
   (a) $x^2 - 7x + 10 = 0$ (b) $x^2 - 16x + 63 = 0$
   (c) $x^2 + 20x + 36 = 0$ (d) $x^2 - 13x + 36 = 0$
   (e) $x^2 + 13x + 40 = 0$

4. Given the following roots, find the equation of the form $Ax^2 + Bx + C = 0$.

(a) $r = \frac{2}{5}$ $s = \frac{3}{5}$

(b) $r = -\frac{2}{3}$ $s = -\frac{4}{3}$

(c) $r = -3$ $s = \frac{1}{7}$

(d) $r = -\frac{1}{4}$ $s = -\frac{1}{2}$

(e) $r = \frac{2}{8}$ $s = -\frac{3}{8}$

(f) $r = -\frac{7}{12}$ $s = \frac{9}{12}$

QUADRATIC FORMULA

Finding the roots of a quadratic equation can be easily computerized by using the quadratic formula. According to this formula, the roots are

$$\frac{-B + \sqrt{B^2 - 4AC}}{2A} \quad \text{or} \quad \frac{-B - \sqrt{B^2 - 4AC}}{2A}$$

Using this formula, the *sum* of the roots is

$$\frac{-B + \sqrt{B^2 - 4AC}}{2A} + \frac{-B - \sqrt{B^2 - 4AC}}{2A}$$

$$= \frac{-B - B + \sqrt{B^2 - 4AC} - \sqrt{B^2 - 4AC}}{2A}$$

$$= -\frac{2B}{2A} = -\frac{B}{A}$$

Again, using the formula, the *product* of the roots is

$$\left(\frac{-B + \sqrt{B^2 - 4AC}}{2A}\right)\left(\frac{-B - \sqrt{B^2 - 4AC}}{2A}\right)$$

$$= \frac{B^2 + B\sqrt{B^2 - 4AC} - B\sqrt{B^2 - 4AC} - (\sqrt{B^2 - 4AC})^2}{4A^2}$$

$$= \frac{B^2 - (B^2 - 4AC)}{4A^2}$$

$$= \frac{B^2 - B^2 + 4AC}{4A^2}$$

$$= \frac{4AC}{4A^2}$$

$$= \frac{C}{A}$$

EXAMPLE

Find the roots of the quadratic equation $15x^2 - 29x - 14 = 0$.

Solution   One root is

$$\frac{-B + \sqrt{B^2 - 4AC}}{2A} = \frac{-(-29) + \sqrt{(-29)^2 - 4(15)(-14)}}{2(15)}$$

$$= \frac{29 + \sqrt{841 + 840}}{30}$$

$$= \frac{29 + \sqrt{1681}}{30}$$

$$= \frac{29 + 41}{30}$$

$$= \frac{70}{30}$$

$$= \frac{7}{3}$$

The other root is

$$\frac{-B - \sqrt{B^2 - 4AC}}{2A}, \qquad \text{which reduces to}$$

$$= \frac{29 - 41}{30}$$

$$= -\frac{12}{30}$$

$$= -\frac{2}{5}$$

The roots are $\frac{7}{3}$ and $-\frac{2}{5}$.

DISCRIMINANT

Because the quadratic formula involves the square root, certain restrictions on the roots will apply.

When $B^2 - 4AC > 0$, if $B^2 - 4AC =$ a perfect square, there are two rational roots. But if $B^2 - 4AC \neq$ a perfect square, there are two irrational roots.

When $B^2 - 4AC = 0$, there is one root.

When $B^2 - 4AC < 0$, there are zero roots since the square root of a negative number does not exist in the real number system.

The numbers of roots in the solution depends on the value of $B^2 - 4AC$. For this reason, $B^2 - 4AC$ is called the *discriminant*.

The flowchart of this concept would be

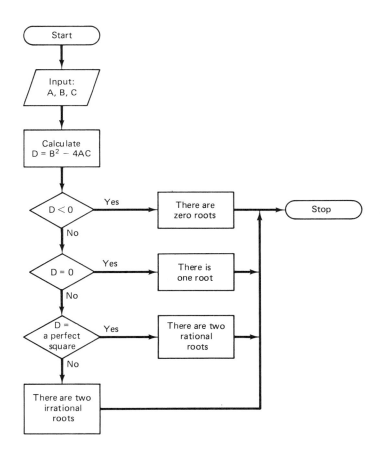

GRAPHING THE QUADRATIC

Quadratic solutions are found by using the equation $Ax^2 + Bx + C = 0$. The roots of the quadratic r and s would be the (x, y) points of $(r, 0)$ and

$(s, 0)$. To graph the equation, it is necessary to find more (x, y) points that satisfy the quadratic.

EXAMPLE 1

Graph $y = 2x^2 + x - 3$.

Solution   First, check to see if any roots exist, and if they do, how many. To find the roots, first solve the discriminant:

$$B^2 - 4AC = 1 - 4(2)(-3)$$
$$= 1 + 24$$
$$= 25$$

25 is greater than 0; therefore, there are two real roots. Using the quadratic formula, one root is

$$\frac{-1 + \sqrt{25}}{4} = \frac{-1 + 5}{4} = 1$$

The other root is

$$\frac{-1 - \sqrt{25}}{4} = \frac{-1 - 5}{4} = -\frac{6}{4} = -\frac{3}{2}$$

If the roots are 1 and $-\frac{3}{2}$, the two points are $(1, 0)$ and $(-\frac{3}{2}, 0)$.
To find other points, pick values for x that are near $-\frac{3}{2}$ and 1:

x	$2x^2 + x - 3 = y$
-2	3
$-\frac{3}{2}$	0
-1	-2
0	-3
$\frac{1}{2}$	-2
1	0
$\frac{3}{2}$	3

The graph is

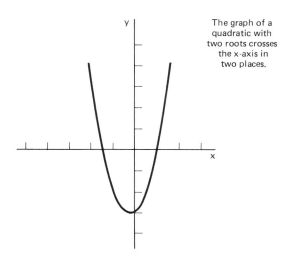

The graph of a quadratic with two roots crosses the x-axis in two places.

This is the general form of a graph of a quadratic equation with two roots.

EXAMPLE 2

Graph $y = 3x^2 + 5x + 7$.

Solution   To find the roots, solve the discriminant.

$$B^2 - 4AC = 25 - 4(3)(7)$$
$$= 25 - 84$$
$$= -59$$

There are no roots, since the discriminant is less than zero. Solve to find other points:

x	y
-2	9
-1.5	6.25
-1	5
$-.5$	5.25
0	7
.5	10.25
1	15

The graph is

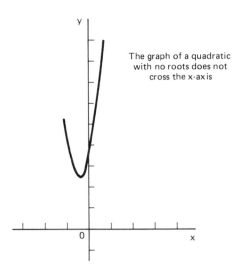

The graph of a quadratic with no roots does not cross the x-axis

EXAMPLE 3

Graph $y = 16x^2 - 24x + 9$.

Solution   To find the roots, solve the discriminant.

$$B^2 - 4AC = (-24)^2 - 4(16)(9)$$

$$= 576 - 576$$

$$= 0 \qquad \text{There is one root.}$$

$$\frac{-(-24) - 0}{2(16)} = \frac{24}{32} = \frac{3}{4}$$

Solve to find the other points.

x	$16x^2 - 24x + 9 = y$
0	9
.25	4
.5	1
.75	0
1	1
1.25	4
1.5	9

The graph is

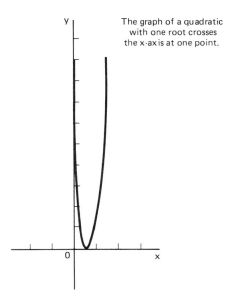

The graph of a quadratic
with one root crosses
the x-axis at one point.

LOWEST POINT

The graph of a quadratic has a lowest point. If a vertical line were drawn through that point, it would divide the graph into mirrored images. To get a more accurate graph, start with the lowest point and take values in equal increments to the right and left.

The standard equation of a quadratic is in two forms:

(1) $$Ax^2 + Bx + C = 0$$

The values of x are in descending order. That means that the C is a coefficient of x^0, which is 1. The equation can be rewritten as

(2) $$Ax^2 + Bx + Cx^0 = 0$$

EXAMPLE 1

With this form of the standard equation, use the following algorithm to find the lowest point.

Step	Given: $y = 3x^2 - 4x + 5$
1. Put the equation in form (2) of the standard equation	$3x^2 - 4x + 5x^0$
2. To find the first term, multiply the exponent by the coefficient	$2(3)$
3. Subtract 1 from the exponent	$6x^{(2-1)} = 6x^1$
4. To find the second term, repeat steps 2 and 3	$1(-4)x^{(1-1)} = -4x^0$ $y = 6x - 4$
5. To find the third term, repeat steps 2 and 3 (*Note:* This term is always 0)	$0(5)x^{(0-1)}$ $y = 6x - 4 + 0$
6. Solve for $y = 0$	$0 = 6x - 4$ $6x = 4$ $x = \frac{4}{6} = \frac{2}{3}$

$x = \frac{2}{3}$. This is the lowest point. Using this value for x, find the corresponding y from the original equation.

$$y = 3x^2 - 4x + 5$$

$$= 3\left(\frac{2}{3}\right)^2 - 4\left(\frac{2}{3}\right) + 5$$

$$= 3\left(\frac{4}{9}\right) - \frac{8}{3} + \frac{15}{3}$$

$$= \frac{4 - 8 + 15}{3}$$

$$= \frac{11}{3}$$

$(\frac{2}{3}, \frac{11}{3})$ is the lowest point. To find symmetric points, use points to the right and left of $\frac{2}{3}$ on the x-axis.

x	y
$\frac{1}{3}$	4
1	4
0	5
$1\frac{1}{3}$	5
$\frac{2}{3}$	$\frac{11}{3}$

The graph is

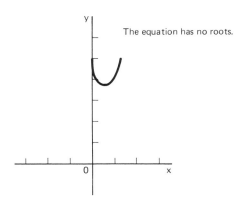

The equation has no roots.

EXAMPLE 2

Find the lowest point and graph $y = 3x^2 - 18x + 25$.

Solution:

$$3x^2 - 18x + 25x^0$$

$$2(3)x^1 - 1(18)x^0 + 0(25)x^{-1}$$

$$0 = 6x - 18$$

$$18 = 6x$$

$$3 = x$$

$$y = 3x^2 - 18x + 25$$

$$= 3(3)^2 - 18(3) + 25$$

$$= 27 - 54 + 25$$

$$= -2$$

The lowest point is $(3, -2)$.
To find the remaining points:

x	y
2	1
4	1
1	10
5	10
3	-2

The graph is

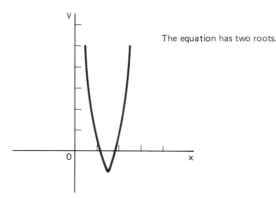

The equation has two roots.

To find the two roots, use the quadratic formula. For the first root,

$$\frac{-(-18) + \sqrt{(-18)^2 - 4(3)(25)}}{2(3)}$$

$$= \frac{18 + \sqrt{324 - 300}}{6}$$

$$= \frac{18 + \sqrt{24}}{6}$$

$$= \text{Approximately } 3.82$$

For the second root,

$$\frac{18 - \sqrt{24}}{6} = \text{approximately } 2.18$$

Both roots are approximate since the $\sqrt{24}$ is irrational.

EXERCISES

1. Find the roots using the quadratic formula.
    (a) $x^2 + 12x + 32 = 0$ (b) $5x^2 + 2x - 3 = 0$
    (c) $6x^2 - 29x + 9 = 0$ (d) $7x^2 - 5x - 2 = 0$
    (e) $9x^2 + 6x - 8 = 0$

2. Find the discriminant and give the number of roots.
    (a) $3x^2 - 17x + 10 = 0$ (b) $9x^2 + 54x + 81 = 0$
    (c) $7x^2 + 13x + 22 = 0$ (d) $4x^2 - 32x + 63 = 0$
    (e) $8x^2 - 11x + 9 = 0$ (f) $36x^2 - 60x + 25 = 0$
    (g) $2x^2 + 35x - 18 = 0$ (h) $2x^2 - 15x + 11 = 0$
    (i) $11x^2 + 20x - 9 = 0$ (j) $16x^2 + 40x + 25 = 0$

3. Find the roots (if they exist).
   (a) $x^2 - 8x + 12 = 0$ (b) $6x^2 + 13x + 5 = 0$
   (c) $8x^2 + 3x + 2 = 0$ (d) $4x^2 - 9x - 3 = 0$
   (e) $7x^2 - 12x + 2 = 0$ (f) $5x^2 + 16x + 11 = 0$
   (g) $4x^2 - 28x + 49 = 0$ (h) $2x^2 + 9x - 13 = 0$
   (i) $11x^2 + 58x - 133 = 0$ (j) $9x^2 - 20x - 5 = 0$

4. In the following equations, if $B = 0$, it is written $Ax^2 + 0x + C = 0$ or $Ax^2 + C = 0$. Find the roots.
   (a) $4x^2 - 25 = 0$ (b) $16x^2 + 121 = 0$
   (c) $121x^2 - 8 = 0$ (d) $25x^2 + 4x = 0$
   (e) $225x^2 - 16x = 0$ (f) $81x^2 - 144 = 0$

5. Find the lowest point.
   (a) $y = 3x^2 - 15x + 10$ (b) $y = 6x^2 - 18x + 12$
   (c) $y = 5x^2 - 11x + 21$ (d) $y = 12x^2 - 48x + 72$
   (e) $y = 9x^2 - 16$ (f) $y = 7x^2 - 35 + 5$
   (g) $y = 3x^2 + 5x$ (h) $y = 8x^2 - 3x + 2$
   (i) $y = 16x^2 + 48x + 20$ (j) $y = 9x^2 - 12x - 8$

6. Find the lowest point and graph using a minimum of six other points.
   (a) $y = x^2 - 7x + 12$ (b) $y = x^2 + 7x + 10$
   (c) $y = x^2 - 16$ (d) $y = x^2 - x - 2$
   (e) $y = 3x^2 - 18x + 7$ (f) $y = 4x^2 + 5x + 1$
   (g) $y = 3x^2 - 8x$ (h) $y = 5x^2 + 3x + 2$
   (i) $y = 4x^2 - x - 3$ (j) $y = 16x^2 - 25$

SUMMARY

The quadratic equation is an equation where the highest exponent is 2. The standard form is $Ax^2 + bx + c = 0$. Every equation has some solution and in a quadratic the roots are the solution: that is, the points at which the values for $y = 0$.

In graphing the equation, the solutions or roots are the equivalent of finding where the graph crosses the x-axis. Here there are three possibilities:

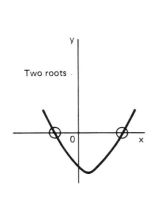

Two roots

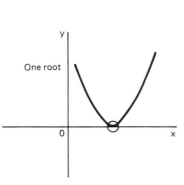

One root

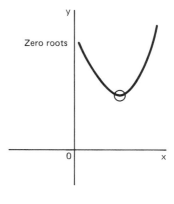

Zero roots

The equation can easily be solved by computer using the quadratic formula. Since the formula involves finding the square root before the formula can be evaluated, the discriminant must be tested. Real roots that can be rational or irrational will have a positive discriminant. A zero discriminant means that the roots are the same. A negative discriminant means that the square root does not exist, or the equation has no solution in the real number system.

In graphing, the graph has a lowest point. A vertical line through this point divides the graph equally since it is symmetrical.

CHAPTER REVIEW

1. The standard form of a quadratic equation is _____.
2. An equation where the highest power is _____ is a quadratic.
3. The roots of a quadratic are the points for which the y values are _____.
4. Quadratic equations can have at most _____ roots.
5. The value $-B/A$ represents the _____ of the roots and C/A represents the _____.
6. The roots of a quadratic can be found using the _____ formula.
7. The quadratic formula is _____.
8. The value of $B^2 - 4AC$ is also called the _____.
9. Fill in the following table:

	Number of real roots
(a) $B^2 - 4AC > 0$	_____
(b) $B^2 - 4AC = 0$	_____
(c) $B^2 - 4AC < 0$	_____

10. The two roots of a quadratic can be _____ or _____.
11. If the discriminant is a perfect square, there are _____ roots and they are _____.
12. The points at which the graph of the quadratic cross the x-axis are the _____ of the equation.
13. If there are no roots, the graph _____ cross the x-axis.
14. A line of symmetry can be drawn through the _____ point.

Linear Equations

chapter 12

Some equations have one variable, such as $3x + 7 = 13$, and have one solution. But an equation such as $Ax^2 + Bx + C = 0$, which is a quadratic, can have as many as two solutions. Certain equations, such as $Ax + By + C = 0$, known as *linear equations*, have *multiple* solutions.

The linear equation has solutions that are known as *ordered pairs*. These ordered pairs form a set of numbers whose graph is a straight line. Because a linear equation represents a straight line, there is an infinite set of points that can satisfy this equality. The equation can give the points of the graph. The reverse is also true—that the graph can give the equation.

Two or more of these equations form a system. A system of linear equations can be represented by lines on a graph. Because these lines can intersect, the system of equations can have a single solution. The solution will be the point where the lines intersect. That point (x, y) will satisfy both (or all) equations in the system.

Since much of computer programming is setting up equations, an understanding of all types of equations and their manipulation is important to the programmer.

LINEAR EQUATIONS

The linear equation has a standard form $Ax + By + C = 0$. This equation is called *linear* because the graph of the equation is a straight line. A line is an infinite set of points, which implies that an infinite set of points can satisfy the equation.

Each point in the set is called an *ordered pair*. They are ordered because they identify a unique point in the Cartesian plane. The point is identified as (x, y).

To graph a line, a minimum of two points is necessary since Euclidean geometry teaches that two points determine a line. However, to ensure that the line is correct, a minimum of three points should be used to draw the graph.

In the standard linear equation form, $Ax + By + C = 0$, the numbers for x can be generated by a loop and the numbers for y can be found by using the x values. In the general equation given x, solve for y by using the following procedure:

$$Ax + By + C = 0$$

$$By = -Ax - C$$

$$y = \frac{-Ax - C}{B} \qquad \text{or} \qquad y = -\frac{A}{B}x - \frac{C}{B}$$

EXAMPLE

$$4x + 3y + 11 = 0$$

Solution   Solve for y first:

$$4x + 3y + 11 = 0$$

$$3y = -4x - 11$$

$$y = -\frac{4}{3}x - \frac{11}{3}$$

Randomly select values for x.

For $x = 1$:
$$y = -\frac{4}{3}(1) - \frac{11}{3}$$

$$= -\frac{4}{3} - \frac{11}{3}$$

$$= -\frac{15}{3}$$

$$= -5$$

One point is $(1, -5)$.

For $x = -2$:
$$y = -\frac{4}{3}(-2) - \frac{11}{3}$$

$$= +\frac{8}{3} - \frac{11}{3}$$

$$= -\frac{3}{3} = -1$$

Another point is $(-2, -1)$.

For $x = 4$:
$$y = -\frac{4}{3}(4) - \frac{11}{3}$$

$$= -\frac{16}{3} - \frac{11}{3}$$

$$= -\frac{27}{3} = -9$$

Another point is $(4, -9)$.

Therefore, three points that satisfy the equation are $(1, -5)$, $(-2, -1)$, and $(4, -9)$. To verify the correctness of these points, substitute them for x and y, and apply them to the sample equation given earlier:

$$(1, -5) \qquad 4(1) + 3(-5) + 11 =$$
$$4 - 15 + 11 = 0$$
$$(-2, -1) \qquad 4(-2) + 3(-1) + 11 =$$
$$-8 - 3 + 11 = 0$$
$$(4, -9) \qquad 4(4) + 3(-9) + 11 =$$
$$16 - 27 + 11 = 0$$

The graph would look like this:

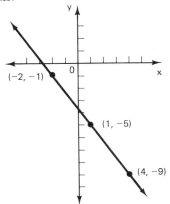

X AND Y INTERCEPTS

A line can cross the x-axis and the y-axis. If the line crosses the x-axis, the point at which it crosses is $(x, 0)$, which is the x-intercept. If the line crosses the y-axis, the point at which it crosses is $(0, y)$, which is the y-intercept.

Using the standard form, the formula for the x-intercept is

$$Ax + By + C = 0$$

$$Ax + B(0) + C = 0$$

$$Ax + C = 0$$

$$Ax = -C$$

$$x = -\frac{C}{A}$$

The x-intercept is $(-C/A, 0)$.

The y-intercept is at $x = 0$. To find x, solve the equation for $x = 0$.

$$A(0) + By + C = 0$$

$$By + C = 0$$

$$By = -C$$

$$y = -\frac{C}{B}$$

The y-intercept is $(0, -C/B)$.

EXAMPLE

Find the x-intercept and the y-intercept for $4x + 5y + 8 = 0$.

Solution   Based on the standard form cited earlier, $A = 4$, $B = 5$, and $C = 8$.

$$x\text{-intercept is:} \quad \left(-\frac{8}{4}, 0\right) \quad \text{or} \quad (-2, 0)$$

$$y\text{-intercept is:} \quad \left(0, -\frac{8}{5}\right)$$

The graph would appear as follows:

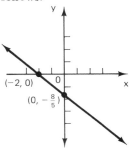

EXERCISES

1. Change to standard form.
   (a) $7x = 3y - 8$ (b) $4y = 2x + 11$
   (c) $y = \frac{3}{5} x - 2$ (d) $x = \frac{7}{8} y + \frac{3}{4}$
   (e) $x - \frac{2}{3} y = 4$ (f) $3x = \frac{2}{3} y + 4$
   (g) $11 = 4x - 3y$ (h) $\frac{7}{5} x + \frac{3}{5} y = 4$
   (i) $3(x + 2) = 4y$ (j) $5(x - \frac{2}{3} y) = 1$

2. Find the y values for the given x values for the equation $2x - 2y - 4 = 0$. Graph the line.

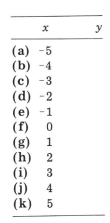

	x	y
(a)	-5	
(b)	-4	
(c)	-3	
(d)	-2	
(e)	-1	
(f)	0	
(g)	1	
(h)	2	
(i)	3	
(j)	4	
(k)	5	

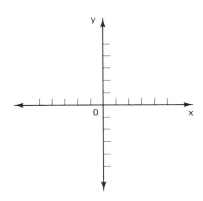

Determine which points satisfy the equation.

3. $4x + 3y + 2 = 0$ A. $(-1, 8)$ B. $(2, -7)$
4. $3x - 2y + 4 = 0$ C. $(11, 10)$ D. $(1, -2)$
5. $5x + y - 3 = 0$ E. $(1, -9)$ F. $(2, 7)$
6. $-2x + 2y - 4 = 0$ G. $(5, -3)$ H. $(2, 9)$
7. $-x + 4y + 12 = 0$ I. $(4, -2)$ J. $(-4, -2)$
8. $8x - 9y + 2 = 0$ K. $(-5, -1)$ L. $(-9, -5)$
9. $7x + 6y - 13 = 0$ M. $(3, -5)$ N. $(0, 4)$
10. $3x - 2y - 19 = 0$ O. $(1, 1)$ P. $(2, -2)$
11. $5x - 3y - 16 = 0$ Q. $(-2, 2)$ R. $(7, 6)$
12. $9x + y - 27 = 0$ S. $(3, 7)$ T. $(2, 2)$

13. Find the x and y intercepts and graph.
    (a) $5x - 3y + 15 = 0$ (b) $2y - 3y - 6 = 0$
    (c) $7x + 2y - 28 = 0$ (d) $3x + 2y - 7 = 0$
    (e) $5x + 4y + 7 = 0$

14. Complete the following table.

	Computer notation	Polish notation
$-C/A$	_____	_____
$-C/B$	_____	_____
$Ax + By + C$	_____	_____
$y = -(A/B)x - C/B$	_____	_____

SLOPE-INTERCEPT FORM

The slope of the line means its "slant" or angle in relation to the x- and y-axis.

Given any two points on the graph, (x_1, y_1) and (x_2, y_2).

$$\text{slope} = \frac{y_2 - y_1}{x_2 - x_1}$$

Given the line $5x + y - 3 = 0$.

x	y
1	-2
0	3
-1	8

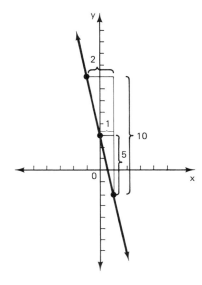

Using $(1, -2)$ and $(-1, 8)$,

$$\text{slope} = \frac{8 - (-2)}{-1 - 1}$$

$$= \frac{8 + 2}{-2}$$

$$= \frac{10}{-2}$$

$$= -5$$

Using $(0, 3)$ and $(-1, 8)$,

$$\text{slope} = \frac{8 - 3}{-1 - 0}$$

$$= \frac{5}{-1}$$

$$= -5$$

On any given line, any two points on that line will always give the same slope.

    $(0, 3)$ is the y-intercept.
    -5 is the slope.

In the original equation

$$5x + y - 3 = 0$$
$$y = -5x + 3$$

                           y-intercept
                           slope

This is the slope-intercept form of the equation.
    Using the standard form,

$$Ax + By + C = 0$$
$$By = -Ax - C$$
$$y = -\frac{A}{B}x - \frac{C}{B}$$

The formula for the slope is $-A/B$. The y-intercept is $(0, -C/B)$.

Using these formulas, it is always possible to find the equation of a line using the graph. Given the slope as m and the y-intercept as $(0, b)$, the equation becomes

$$y = mx + b$$

EXAMPLE 1

Using the following equation, find the slope and y-intercpet.

$$2x + 3y - 8 = 0$$

Solution:

1. Using the formula with

$$A = 2$$
$$B = 3$$
$$C = -8$$

then

$$\text{slope} = -\frac{A}{B} = -\frac{2}{3}$$

and

$$y\text{-intercept} = \left(0, -\frac{C}{B}\right) = \left(0, -\frac{-8}{3}\right)$$
$$= \left(0, \frac{8}{3}\right)$$

2. Using the slope-intercept form,

$$3y = -2x + 8$$
$$y = -\frac{2}{3}x + \frac{8}{3}$$

Thus

$$\text{slope} = -\frac{2}{3}$$
$$y\text{-intercept} = \frac{8}{3}$$

EXAMPLE 2

Given the following graph, find the standard form equation of the line.

Solution:

1. Find the y-intercept: $(0, 7)$.

2. Pick any two points: $(4, 3)$ and $(1, 6)$.

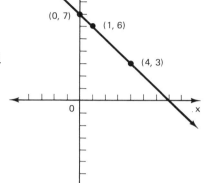

$$\text{Slope} = \frac{6 - 3}{1 - 4} = \frac{3}{-3} = -1$$

The equation is therefore

$$y = -1x + 7$$

which in standard form is

$$x + y - 7 = 0$$

EXAMPLE 3

Given two points, find the standard form equation for the line: $(-2, 1)$ and $(2, 3)$.

Solution:

1. Find the slope.

$$\text{Slope} = \frac{3 - 1}{2 - (-2)} = \frac{2}{4} = \frac{1}{2}$$

2. Use the slope-intercept form with slope $y = \frac{1}{2}x + b$ and choose one of the points, such as $(2, 3)$. Then solve the equation for b:

$$3 = \frac{1}{2}(2) + b$$

$$= 1 + b$$

$$2 = b$$

The equation is

$$y = \frac{1}{2}x + 2$$

Change to standard form by multiplying by 2:

$$(2)y = (2) \frac{1}{2} x + (2)2$$

$$2y = x + 4$$

$$-x + 2y - 4 = 0$$

Certain equations can be said to have no slope or a slope of 0. The slope intercept form of the standard equation is

$$y = -\frac{A}{B} x - \frac{C}{B}$$

1. If $B = 0$, then finding the slope of y-intercept would involve division by 0. In this case, the slope does not exist; neither does the y-intercept. The standard equation would be $Ax + C = 0$. This is an equation with no y term.

2. If $A = 0$, then the equation is

$$y = -\frac{0}{B} x - \frac{C}{B}$$

$$= 0x - \frac{C}{B}$$

The slope is 0.

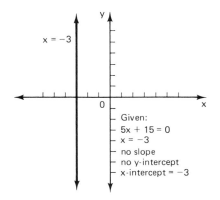

Given:
5x + 15 = 0
x = −3
no slope
no y-intercept
x-intercept = −3

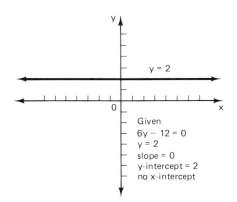

Given
6y − 12 = 0
y = 2
slope = 0
y-intercept = 2
no x-intercept

EXERCISES

1. Change the equations to slope-intercept form and complete the table.

	Equation	Slope-intercept form	Slope	y-intercept
(a)	$5x - 3y + 7 = 0$	_____	_____	_____
(b)	$8x + 2y - 12 = 0$	_____	_____	_____
(c)	$3x + 2y - 15 = 0$	_____	_____	_____
(d)	$7x + y - 14 = 0$	_____	_____	_____
(e)	$11x - 2y - 44 = 0$	_____	_____	_____
(f)	$8x + 3y + 24 = 0$	_____	_____	_____
(g)	$4x = 7y - 28$	_____	_____	_____
(h)	$x + 6y = 24$	_____	_____	_____
(i)	$3x - 27 = 9y$	_____	_____	_____
(j)	$5x + 7y = 35$	_____	_____	_____

2. Find the standard equation of graphs A, B, C, D, E, and F.

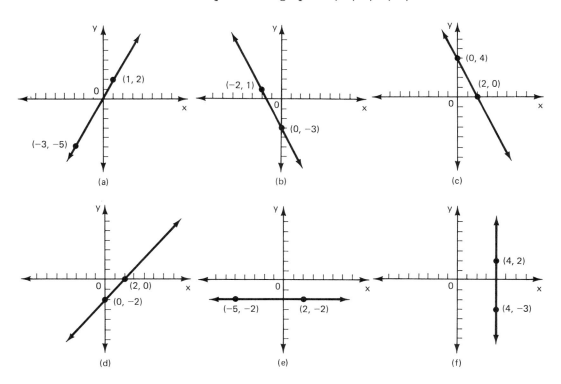

3. Find the standard equation for each set of points.
   (a) (−2, 3) (4, −1)   **(b)** (2, 6) (−1, 9)
   (c) (1, −3) (7, 2)   **(d)** (−2, 1) (3, 3)
   (e) (−3, −1) (1, 6)   **(f)** (−3, 6) (4, 2)
   (g) (−5, 2) (2, −3)   **(h)** (3, 5) (−3, −4)
   (i) (2, 1) (1, 10)   **(j)** (8, 2) (−3, −1)
4. Find the slopes of the lines through each set of points.
5. Find the slope-intercept equation for each set of points.

SYSTEM OF EQUATIONS

A set of two linear equations gives two lines on a graph. Two lines can cross at one point. This point (x, y) will satisfy both equations. This point is said to be the *solution* of the system of equations.

If the two lines cross at one point, the solution is the following:

$$(1) \quad Ax + By + C = 0$$

$$(2) \quad Dx + Ey + F = 0$$

To solve for y, multiply (1) by D and (2) by $-A$:

$$ADx + BDy + CD = 0$$

$$-ADx - AEy - AF = 0$$

Adding the two equations results in

$$(BD - AE)y + (CD - AF) = 0$$

$$(BD - AE)y = -(CD - AF)$$

$$y = \frac{-(CD - AF)}{BD - AE}$$

To solve for x, multiply (1) by $-E$ and (2) by B:

$$-AEx - BEy - CE = 0$$

$$BDx + BEy + BF = 0$$

Adding the two equations results in

$$(BD - AE)x + (BF - CE) = 0$$

$$(BD - AE)x = -(BF - CE)$$

$$x = \frac{-(BF - CE)}{BD - AE}$$

EXAMPLE 1

Find the solution for the following system of equations:

$$4x + 2y - 16 = 0$$

$$-6x + 2y - 6 = 0$$

Solution:

$$A = 4 \qquad D = -6$$

$$B = 2 \qquad E = 2$$

$$C = -16 \qquad F = -6$$

$$y = \frac{-[-16(-6) - 4(-6)]}{2(-6) - 4(2)}$$

$$= \frac{-(96 + 24)}{-12 - 8}$$

$$= \frac{-120}{-20}$$

$$= 6$$

$$x = \frac{-[2(-6) - (-16)(2)]}{-20}$$

$$= \frac{-(-12 + 32)}{-20}$$

$$= \frac{-20}{-20}$$

$$= 1$$

To verify that (1, 6) is the solution:

$$4x + 2y - 16 = 0 \qquad\qquad -6x + 2y - 6 = 0$$

$$4(1) + 2(6) - 16 = 0 \qquad\qquad -6(1) + 2(6) - 6 = 0$$

$$4 + 12 - 16 = 0 \qquad\qquad -6 + 12 - 6 = 0$$

$$16 - 16 = 0 \qquad\qquad 0 = 0$$

$$0 = 0$$

The solution can be seen by graphing:

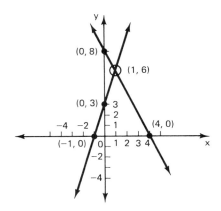

Some equations never cross, which means that the equation lines are parallel. In this case, the system is said to have no solution.

EXAMPLE 2

Given the system

$$3x + y - 3 = 0$$

$$6x + 2y + 4 = 0$$

$$A = 3 \qquad D = 6$$

$$B = 1 \qquad E = 2$$

$$C = -3 \qquad F = 4$$

$$BD - AE = (1)(6) - (3)(2) = 6 - 6 = 0$$

The graph is as follows:

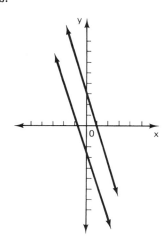

Since division by 0 is not allowed, no solution exists.

Look at the slope-intercept forms of the equations

$$(1) \quad y = -3x + 3$$
$$(2) \quad y = -3x - 2$$

Both equations have the same slope but different y-intercepts. To find the solution to a system of equations, use the following flowchart method:

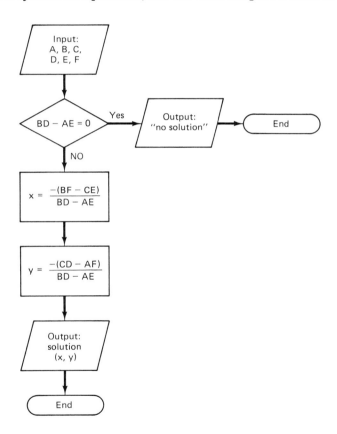

EXERCISES

1. Determine if a solution exists, and if it does, find the solution.
   (a) $2x + 3y = 4$
       $3y - 6x = 12$
     (b) $2x + 2y = 16$
       $5x - 2y = -2$
   (c) $5x + 3y + 15 = 0$
       $2x - 3y + 27 = 0$
     (d) $-3x - 2y = -4$
       $5x - 4y = 36$

(e) $3x - 7y - 14 = 0$
$6x - 14y + 42 = 0$

(f) $3x - 27 = 0$
$4y + 8 = 0$

(g) $y = \frac{2}{3} x + 9$
$y = -\frac{3}{2} x + 2$

(h) $2y = x + 3$
$3y = -5x - 15$

(i) $2x - 3y + 5 = 0$
$y = \frac{2}{3} x + \frac{8}{3}$

(j) $3y - 12 = 0$
$5y + 25 = 0$

Match the systems with their solutions.

2. $2x - y + 10 = 0$
$x - 2y + 12 = 0$

3. $3x + 2y = 20$
$2y - x = 12$

4. $y = \frac{4}{3} x + 10$
$y = 2x + 10$

5. $5x - 8y = 63$
$4x + 3y = -6$

6. $2y - 5x - 10 = 0$
$5y = -7x - 53$

7. $y = \frac{5}{2} x + 5$
$y = \frac{1}{2} x + 6$

8. $11x - 6y - 56 = 0$
$3x - 10y + 60 = 0$

9. $y = -\frac{6}{7} x - 2$
$y = \frac{2}{7} x + 6$

10. $-6y = 11x - 29$
$6x + 7y + 14 = 0$

A. $(3, -6)$
B. $(\frac{1}{2}, \frac{25}{4})$
C. $(-\frac{8}{3}, \frac{14}{3})$
D. $(-7, 4)$
E. $(0, 10)$
F. $(10, 9)$
G. $(1, -3)$
H. $(2, 7)$
I. $(-4, 13)$
J. $(-4, -5)$
K. $(7, -8)$

11. Write in computer notation.

(a) $BD - AE$

(b) $x = \dfrac{-(BF - CE)}{BD - AE}$

(c) $y = \dfrac{-(CD - AF)}{BD - AE}$

SUMMARY———————————————————————

Equations are of many different types. Linear equations are equations of the form

$$Ax + By + C = 0$$

An equation is called a linear equation because its graph is a straight line. The line is a set of infinite points (x, y). These points are the set of ordered pairs for which the given equation is true.

The linear equation is graphed on the Cartesian plane, with an x-axis and y-axis. Every line has certain identifiable points. The x-intercept is $(-C/A, 0)$, and the y-intercept is $(0, -C/B)$. If $A = 0$, there is no x-intercept, and the equation becomes $By + C = 0$. In this case, for any value of x, the value of y is always the same. Similarly, if $B = 0$, there is no y-intercept and the equation becomes $Ax + C = 0$. The value of x is constant for any value of y.

The slant of the line is known as the slope. The slope is defined as

$$\frac{y_2 - y_1}{x_2 - x_1}$$

The slope-intercept form of the equation is found by changing the standard equation into

$$y = -\frac{A}{B}x - \frac{C}{B}$$

Combining two linear equations gives a system of equations. These equations can have a solution that is an ordered pair (x, y), which satisfies both equations. Graphically, this point is where both lines intersect. Systems of equations can have no solution. In this case, graphically, the lines never cross, or are the same line.

CHAPTER REVIEW

1. $Ax + By + C = 0$ is the standard form of a _____ equation.
2. The graph of $Ax + By + C = 0$ is a _____ .
3. The points (x, y) that satisfy the equation $Ax + By + C = 0$ are called

   _____ .
4. To graph a line requires a minimum of _____ points.
5. If three points satisfy the same equation, the three points _____

   _____ .
6. The point $(-C/A, 0)$ is the _____ , and the point $(0, -C/B)$ is the _____ .
7. The slant of a graphed line is known as the _____ of the line.

8. In the equation

$$y = -\frac{A}{B} x - \frac{C}{B}$$

the slope is _____ and the y-intercept is _____ .

9. An equation of the form $Ax + C = 0$ has _____ slope.

10. An equation of the form $By + C = 0$ has _____ slope.

11. A system of linear equations having one solution means _____
_____ .

12. If two lines in a system are parallel, then _____
_____ .

13. Two equations having the same slopes but different y-intercepts will have _____ solutions.

14. In the graph of a system of equations, the solution is the point _____
_____ .

Linear Programming

Frequently, linear equations are used to represent real-life situations. The question that is usually asked about these situations is: "What is the maximum?" or "What is the minimum?" Linear programming is the mathematical way of finding this answer.

The maximum or minimum is the (x, y) point which gives the largest or smallest value for a linear equation. The linear programming method involves setting up constraints using linear inequalities. These graph into a polygon with vertices. The vertices are (x, y) points which give the maximum and minimum values for a linear equation.

LINEAR INEQUALITIES

The inequality conditions are:

$>$ Greater than
$<$ Less than
$\geqslant$ Greater than or equal to
$\leqslant$ Less than or equal to

The linear inequality is a conditional linear equation. A linear inequality is of the form

$$6x + 3y \leqslant 24$$

The points in the truth set are the (x, y) values that make the equation true. Given the following list, find the ordered pairs in the truth set:

(x, y)	$6x + 3y \leqslant 24$	
$(1, 10)$	$6(1) + 3(10) = 36 \leqslant 24$	False
$(1, 8)$	$6(1) + 3(8) = 30 \leqslant 24$	False
$(1, 6)$	$6(1) + 3(6) = 24 \leqslant 24$	True
$(1, 4)$	$6(1) + 3(4) = 18 \leqslant 24$	True
$(2, 3)$	$6(2) + 3(3) = 21 \leqslant 24$	True
$(2, 5)$	$6(2) + 3(5) = 27 \leqslant 24$	False
$(2, 10)$	$6(2) + 3(10) = 42 \leqslant 24$	False

The values in the truth set are $(1, 6)$, $(1, 4)$, $(2, 3)$.

There is an inequality for each value of x. For example, by manipulation the equation becomes

$$3y \leqslant -6x + 24$$
$$y \leqslant -2x + 8$$

If $x = 1$, then $y \leqslant -2(1) + 8$, or $y \leqslant 6$. The truth set for $x = 1$ is

$$(1, 6), (1, 5), (1, 4), \ldots$$

If $x = 2$, then $y \leqslant -2(2) + 8$, or $y \leqslant 4$. The truth set for $x = 2$ is

$$(2, 4), (2, 3), (2, 2), \ldots$$

GRAPHING INEQUALITIES

The graph for the set of points of a linear equation is a line. The graph for the set of points of a linear inequality is a *half-plane*.

EXAMPLE 1

Graph the inequality $6x + 3y \leqslant 24$.

Solution:

For $x = 1$:	$y \leqslant 6$		For $x = 0$:	$y \leqslant 8$
$x = 2$:	$y \leqslant 4$		$x = -1$:	$y \leqslant 10$
For $x = 3$:	$y \leqslant 2$		For $x = -2$:	$y \leqslant 12$
$x = 4$:	$y \leqslant 0$		$x = -3$:	$y \leqslant 14$

Graphing these points yields the following

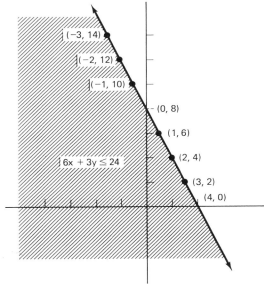

The shaded side of the graph represents a half-plane. The dividing line is the line $6x + 3y = 24$. All the points on the line, and to the left of the line, satisfy the inequality $6x + 3y \leqslant 24$.

EXAMPLE 2

Graph the inequality $3x - 2y \geqslant 6$.

Solution:

Step 1: Find the points for *arbitrary values of x*:

$$x = -2: \quad -6 - 2y \geqslant 6$$
$$-2y \geqslant 12$$
$$y \leqslant -6$$

$$x = -1: \qquad -3 - 2y \geqslant 6$$
$$-2y \geqslant 9$$
$$y \leqslant -4.5$$

$$x = 0: \qquad -2y \geqslant 6$$
$$y \leqslant -3$$

$$x = 1: \qquad 3 - 2y \geqslant 6$$
$$-2y \geqslant 3$$
$$y \leqslant -1.5$$

$$x = 2: \qquad 6 - 2y \geqslant 6$$
$$-2y \geqslant 0$$
$$y \leqslant 0$$

Step 2: Graph the y inequality for each value of x:

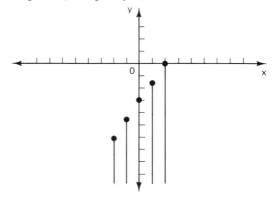

Step 3: Connect the (x, y) coordinates with a line:

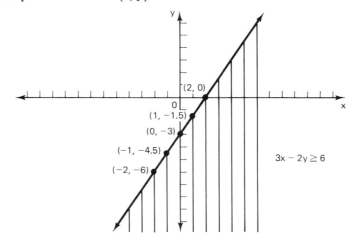

EXAMPLE 3

Graph the inequality $x + y > 0$.

Solution:

    Step 1: Find the points for arbitrary values of x:

$$
\begin{aligned}
x = -2: &\quad -2 + y > 0 \\
        &\quad\quad\quad\ y > 2 \\
x = -1: &\quad -1 + y > 0 \\
        &\quad\quad\quad\ y > 1 \\
x = 0: &\quad\ 0 + y > 0 \\
        &\quad\quad\quad\ y > 0 \\
x = 1: &\quad\ 1 + y > 0 \\
        &\quad\quad\quad\ y > -1 \\
x = 2: &\quad\ 2 + y > 0 \\
        &\quad\quad\quad\ y > -2 \\
x = 3: &\quad\ 3 + y > 0 \\
        &\quad\quad\quad\ y > -3
\end{aligned}
$$

    Step 2: Graph the y inequalities for each value of x:

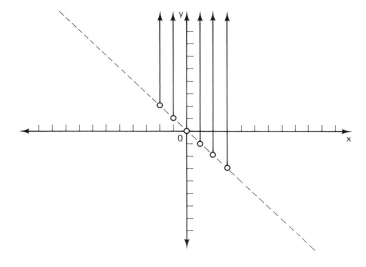

The open circle is used around the value for y because the inequality is only "greater than."

Step 3: As shown in Step 2, the open circles are connected with a dashed line. Every point to the right of the line—not including the points on the line—solve the inequality.

EXERCISES

1. Given the ordered pairs

$$(1, 1), (1, 2), (1, 3), (1, 4), (1, 5)$$
$$(2, 1), (2, 2), (2, 3), (2, 4), (2, 5)$$
$$(3, 1), (3, 2), (3, 3), (3, 4), (3, 5)$$
$$(4, 1), (4, 2), (4, 3), (4, 4), (4, 5)$$
$$(5, 1), (5, 2), (5, 3), (5, 4), (5, 5)$$

determine which are in the truth sets of these inequalities.

(a) $3x + 2y < 12$ (b) $3x + 2y > 12$
(c) $3x + 2y \leqslant 12$ (d) $3x + 2y \geqslant 12$
(e) $x - 5y \leqslant 10$ (f) $x - 5y \geqslant 10$
(g) $x + 2y \leqslant 18$ (h) $x + 2y \geqslant 18$
(i) $2x + y \leqslant 0$ (j) $2x + y \geqslant 0$

Match the region shaded with the corresponding inequality shown following the graphs.

2.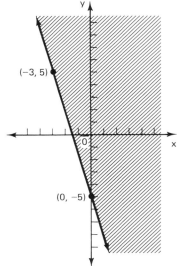

3.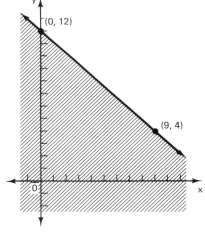

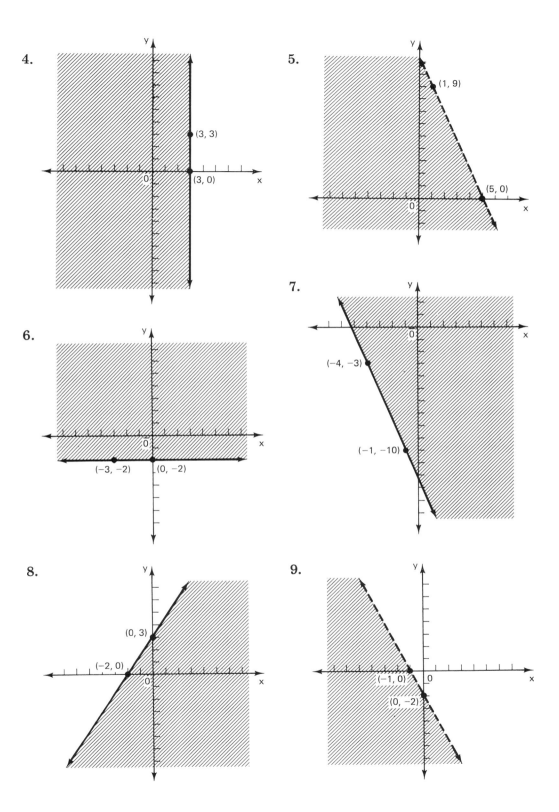

4.

(3, 3)

0

(3, 0)

x

y

5.

(1, 9)

(5, 0)

0

x

y

6.

0

x

y

(−3, −2)   (0, −2)

7.

0

x

y

(−4, −3)

(−1, −10)

8.

y

(0, 3)

(−2, 0)

0

x

9.

y

(−1, 0)   0

x

(0, −2)

10.

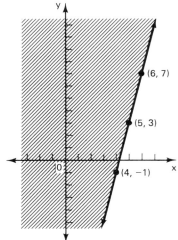

A. $9x + 4y < 45$

B. $x \leqslant 3$

C. $2x + y \leqslant -2$

D. $10x + 3y \leqslant -15$

E. $4x - y \leqslant 17$

F. $7x + 3y \geqslant -37$

G. $3x - 2y \geqslant -6$

H. $8x + 9y \leqslant 108$

I. $2x + y < -2$

J. $y \geqslant -2$

K. $9x + 4y > 45$

11. Graph.

(a) $4x + y \geqslant 15$

(b) $6x - 5y \leqslant 30$

(c) $2x + 3y \geqslant -12$

(d) $3x - 4y \leqslant 24$

(e) $x - 3y > 9$

(f) $5x + 2y < 20$

SYSTEMS OF INEQUALITIES

Just as a set of two or more equations forms a system of equations, so does a set of two or more inequalities form a system of inequalities.

EXAMPLE 1

$$2x + y \geqslant 4$$

$$x + 3y \leqslant 9$$

Solution   To solve, both should be graphed on the same axis.

Step 1:

$2x + y \geqslant 4$		$x + 3y \leqslant 9$	
$x = -2$:	$-4 + y \geqslant 4$	$x = -6$:	$-6 + 3y \leqslant 9$
	$y \geqslant 8$		$3y \leqslant 15$
			$y \leqslant 5$
$x = -1$:	$-2 + y \geqslant 4$		
	$y \geqslant 6$		
$x = 0$:	$0 + y \geqslant 4$	$x = -3$:	$-3 + 3y \leqslant 9$
	$y \geqslant 4$		$3y \leqslant 12$
			$y \leqslant 4$
$x = 1$:	$2 + y \geqslant 4$	$x = 0$:	$3y \leqslant 9$
	$y \geqslant 2$		$y \leqslant 3$
$x = 2$:	$4 + y \geqslant 4$	$x = 3$:	$3 + 3y \leqslant 9$
	$y \geqslant 0$		$3y \leqslant 6$
			$y \leqslant 2$
		$x = 6$:	$6 + 3y \leqslant 9$
			$3y \leqslant 3$
			$y \leqslant 1$

Step 2: Graph both on the same axis, using vertical lines (|||||) for $2x + y \geqslant 4$, and horizontal lines ($\equiv$) for $x + 3y \leqslant 9$.

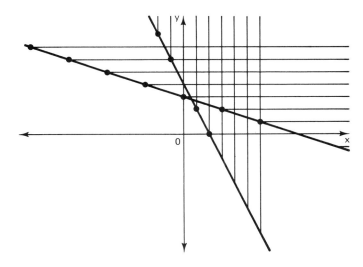

The solution is where the two half-planes intersect.

EXAMPLE 2

Find the solution of the following system:

$$x \geqslant 0$$

$$5x - 2y \geqslant -10$$

$$x + y \leqslant 9$$

Solution:

Step 1:

$5x - 2y \geqslant -10$		$x + y \leqslant 9$	
$x = -2$:	$\begin{aligned}-10 - 2y &\geqslant -10\\ -2y &\geqslant 0\\ y &\leqslant 0\end{aligned}$	$x = 3$:	$\begin{aligned}3 + y &\leqslant 9\\ y &\leqslant 6\end{aligned}$
$x = 0$:	$\begin{aligned}-2y &\geqslant -10\\ y &\leqslant 5\end{aligned}$	$x = 2$:	$\begin{aligned}2 + y &\leqslant 9\\ y &\leqslant 7\end{aligned}$
$x = -4$:	$\begin{aligned}-20 - 2y &\geqslant -10\\ -2y &\geqslant 10\\ y &\leqslant -5\end{aligned}$	$x = 1$:	$\begin{aligned}1 + y &\leqslant 9\\ y &\leqslant 8\end{aligned}$

Step 2: Graph each on the same axis: $x \geqslant 0$ (||||); $5x - 2y \geqslant -10$ ($\equiv$); $x + y \leqslant 9$ (/////).

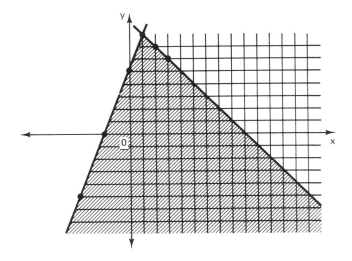

The solution is the intersection of the three half-planes.

POLYGONS DETERMINED BY LINEAR SYSTEMS

The intersection of inequalities can result in a polygon. A *polygon* is formed by lines intersecting to form a closed plane. The points at which the lines intersect are the vertices.

EXAMPLE 1

Given the system

$$8x + 6y \leqslant 48$$

$$x - y \geqslant -1$$

$$2x - 9y \leqslant 54$$

find the vertices of the polygon formed by the intersection of the three half-planes.

Solution:

Step 1:   Determine each half-plane separately:

$8x + 6y \leqslant 48$	$x - y \geqslant -1$	$2x - 9y \leqslant 54$
For x = 3:	*For x = 2:*	*For x = 9:*
$24 + 6y \leqslant 48$	$2 - y \geqslant -1$	$18 - 9y \leqslant 54$
$6y \leqslant 24$	$-y \geqslant -3$	$-9y \leqslant 36$
$y \leqslant 4$	$y \leqslant 3$	$y \geqslant -4$
For x = 0:	*For x = -1:*	*For x = 0:*
$0 + 6y \leqslant 48$	$-1 - y \geqslant -1$	$0 - 9y \leqslant 54$
$y \leqslant 8$	$-y \geqslant 0$	$y \geqslant -6$
	$y \leqslant 0$	

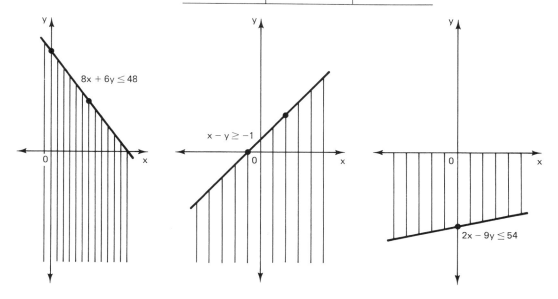

Step 2: Graph the planes on the same axis.

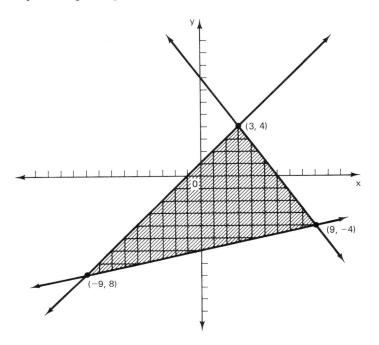

Step 3: Determine the vertices $(-9, 8)$, $(9, -4)$, and $(3, 4)$ of the polygon.

EXAMPLE 2

Find the vertices of the polygon formed by the following system of inequalities:

$$x \geqslant 0$$

$$y \geqslant 0$$

$$x + 5y \leqslant 70$$

$$13x + 5y \leqslant 130$$

Solution:

Step 1: Determine graphs for each inequality:

$x + 5y \leqslant 70$	$13x + 5y \leqslant 130$
For x = 10:	*For x = 5:*
$10 + 5y \leqslant 70$	$65 + 5y \leqslant 130$
$5y \leqslant 60$	$5y \leqslant 65$
$y \leqslant 12$	$y \leqslant 13$

$x + 5y \leqslant 70$	$13x + 5y \leqslant 130$
For x = 0:	*For x = 10:*
$0 + 5y \leqslant 70$ $y \leqslant 14$	$130 + 5y \leqslant 130$ $5y \leqslant 0$ $y \leqslant 0$
$x \geqslant 0$	$y \geqslant 0$
All points to the right of the *y*-axis	All points above the *x*-axis

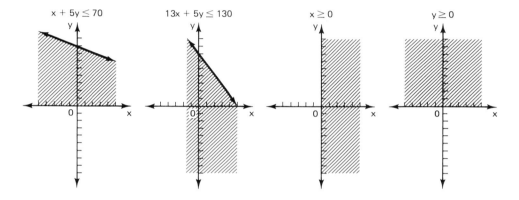

Step 2: Graph all inequalities on the same axis:

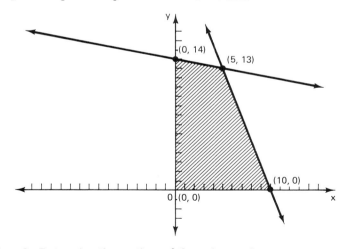

Step 3: Determine the vertices of the polygon above

$$(0, 0), (0, 14), (10, 0), (5, 13)$$

FINDING MINIMUM AND MAXIMUM

To find the minimum and maximum points in linear programming, two factors are needed:

1. A general linear equation, where the slope is given, but where the y-intercept is variable.
2. A polygon which is determined by a system of inequalities.

A linear equation such as $F = 2x + y$ is a general linear equation with a specified slope but with a y-intercept that varies. $F = 2x + y$, in slope-intercept form, becomes $y = -2x + F$. The slope is -2; the y-intercept is the variable F.

Finding a minimum or a maximum means there are constraints placed on the linear equation. These constraints are a system of inequalities which form a polygon. The maximum or minimum of the equation is determined by these constraints.

EXAMPLE 1

Given the system

$$x \geqslant 0$$
$$y \geqslant 0$$
$$x + 5y \leqslant 70$$
$$13x + 5y \leqslant 130$$

find the maximum or minimum values of $F = 2x + y$.

Solution   This graph, shown in Example 2 of the preceding section, can be used to determine the maximum and minimum of the equation $F = 2x + y$.

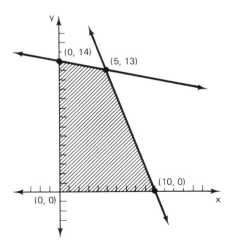

If graphed through $(0, 0)$, $(10, 0)$, $(0, 14)$, and $(5, 13)$, the values are:

For $(0, 0)$:	$F = 0$
For $(10, 0)$:	$F = 20$
For $(0, 14)$:	$F = 14$
For $(5, 13)$:	$F = 26$

When graphed, these equations form a series of parallel lines.

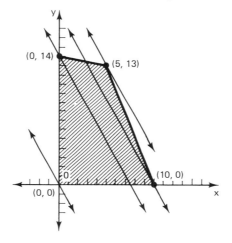

The values of F are:

(x, y)	$F = 2x + y$	
$(0, 0)$	0	← minimum
$(0, 14)$	14	
$(10, 0)$	20	
$(5, 13)$	26	← maximum

These determine that the minimum value of F is 0 and the maximum value is 26.

EXAMPLE 2

Determine the minimum and maximum values of

$$F = 5x + y$$

The following inequalities are given as constraints:

$$7x + 4y \geqslant 28$$

$$-6x + 5y \leqslant 35$$

$$19x - 6y \leqslant 76$$

Solution   Graphing the inequalities gives the vertices of the polygon as $(0, 7)$, $(4, 0)$, and $(10, 19)$.

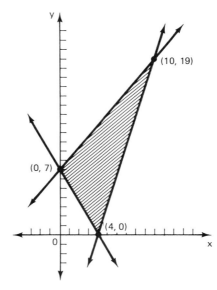

(0, 7) $F = 5(0) + 7 = 7$ minimum

(4, 0) $F = 5(4) + 0 = 20$

(10, 19) $F = 5(10) + 19 = 69$ maximum

EXERCISES

Match the region shaded with the corresponding system of inequalities shown following the graphs.

1.

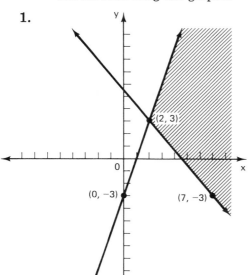

2.

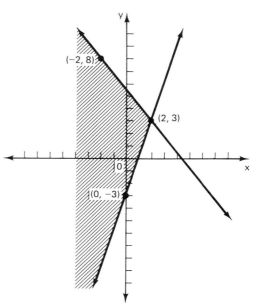

3.

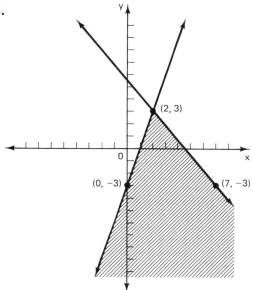

4.

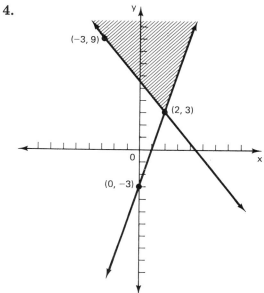

A. $6x + 5y \geqslant 27$ **B.** $6x + 5y \geqslant 27$ **C.** $6x + 5y \leqslant 27$ **D.** $6x + 5y \leqslant 27$
   $3x - y \geqslant 3$ $3x - y \leqslant 3$ $3x - y \geqslant 3$ $3x - y \leqslant 3$

5. Graph the solution.
   (a) $x \geqslant 2$ **(b)** $x \geqslant 0$
       $y \leqslant 0$ $4x - 3y \leqslant 9$
   (c) $y \geqslant 0$ **(d)** $2x + 3y \geqslant 18$
       $2x + y \leqslant 7$ $4x + 3y \leqslant 30$
   (e) $4x + 5y \leqslant 7$ **(f)** $x - y \geqslant -1$
       $7x - 3y \geqslant -23$ $2x + 7y \leqslant 23$

6. Identify the vertices.
   (a)

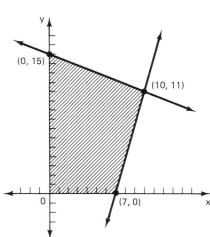

(b)

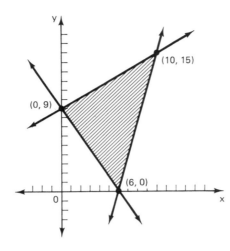

7. Find the vertices of the polygons determined by each system of inequalities.

(a) $x + y > 7$
$6x - 11y \geqslant -77$
$13x - 4y \leqslant 91$

(b) $y \geqslant 0$
$8x - 9y \geqslant 0$
$8x + 5y \leqslant 112$

(c) $3x - 11y \geqslant -27$
$y \geqslant 0$
$3x - 2y \geqslant 0$
$2x - y \leqslant 20$

(d) $x \geqslant 0$
$y \geqslant 0$
$6x - 7y \geqslant -4$
$5x + 2y \leqslant 55$

8. Find the minimum and maximum values for $5x + 7y = P$ for the systems in Exercise 7.

SUMMARY

In this chapter the following ideas were developed and illustrated:

1. A conditional linear equation is called an inequality.

2. A truth set is the set of value(s) that satisfy the inequalities.

3. The values in a truth set of an inequality are the (x, y) points whose graph is a half-plane.

4. Two inequalities have as their solution the intersection of two half-planes.

5. Linear programming is a method for determining the maximum and minimum values by the following:

(a) A system of linear inequalities forming the constraints of a mathematical situation. The area enclosed by the graph lines form a polygon.

(b) A linear equation with constant slope and a variable y-intercept.

(c) Using the vertices of the polygon in the linear equation.

CHAPTER REVIEW _____

1. The linear inequality is the _____ linear equation.

2. The solution of an inequality is a set of _____ .

3. The graph of $3x + 4y \leqslant 12$ is a _____ plane whose dividing line is _____ .

4. A system of inequalities has a graph that is formed by the _____ _____ .

5. If three or more inequalities intersect, they can intersect in a _____ _____ .

6. The closed plane formed by the intersection of the lines and half-planes has points where the lines intersect, called _____ .

7. The points of the intersecting lines can be used to find _____ and _____ solution to a linear equation.

8. The inequalities are _____ placed on the linear equation.

9. The system using the inequalities and the linear equation for maximum or minimum solution is called _____ .

Functions

In mathematics, a *function* represents a special relationship between two numbers. The two numbers are identified as x and $f(x)$. $f(x)$ is the function of x, or a value that is dependent on x.

Programming languages also have functions. These functions will be mathematical expressions which are intrinsic to the language. Functions can also be defined by the programmer and are usually mathematical formulas. In either case, understanding functions mathematically helps in the understanding and use of functions in programming.

BASIC DEFINITIONS OF FUNCTIONS

Given any set of ordered pairs

$$\{(2, 4), (3, 6), (4, 8), (5, 10), (6, 12)\}$$

the ordered pairs form a relation. *If, for each value of x in the ordered pairs, there is only one value for y, the relation is said to be a function.* For example, given the following relations:

$$A = \{(1, 3), (2, 5), (4, 7), (6, 9)\}$$
$$B = \{(1, 2), (2, 3), (2, 4), (3, 1)\}$$
$$C = \{(5, 3), (6, 3), (7, 3), (8, 3)\}$$

A is a function since *each value of x has only one value for y.*

B is *not* a function since when x is 2, y is 3 or 4. x has more than one value for y.

C is a function because for each value of x there is only one value for y. *The y values can be repeated but the x values cannot.*

The values of x in a function are the *domain* of the function.

The values of y in a function are the *range* of the function.

In A the domain is $\{1, 2, 4, 6\}$; the range is $\{3, 5, 7, 9\}$.

In C the domain is $\{5, 6, 7, 8\}$; the range is $\{3\}$.

THE FUNCTION $f(x)$

The function $f(x)$ is defined by an equation. For example, given a function

$$f(x) = 2x + 5$$

if the domain is the set of numbers $\{-3, -2, -1, 0, 1, 2, 3\}$, then to find the range, evaluate the function for each value in the domain:

$$f(-3) = 2(-3) + 5 = -1$$
$$f(-2) = 2(-2) + 5 = 1$$
$$f(-1) = 2(-1) + 5 = 3$$
$$f(0) = 2(0) + 5 = 5$$
$$f(1) = 2(1) + 5 = 7$$
$$f(2) = 2(2) + 5 = 9$$
$$f(3) = 2(3) + 5 = 11$$

The range of the function is $\{-1, 1, 3, 5, 7, 9, 11\}$.

The function $f(x)$ can be evaluated for any set of numbers or variables.

For example, if $f(x) = x^2 - 6$, then:

1. Find $f(\frac{1}{2})$.

    $$f(\tfrac{1}{2}) = (\tfrac{1}{2})^2 - 6$$
    $$= \tfrac{1}{4} - 6 = -5\tfrac{3}{4}$$

2. Find $f(.1)$.

    $$f(.1) = (.1)^2 - 6$$
    $$= .01 - 6 = -5.99$$

3. Find $f(a)$.

    $$f(a) = (a)^2 - 6$$
    $$= a^2 - 6$$

4. Find $f(b + 1)$.

    $$f(b + 1) = (b + 1)^2 - 6$$
    $$= b^2 + 2b + 1 - 6$$
    $$= b^2 + 2b - 5$$

5. Find $f(z/2)$.

    $$f\left(\frac{z}{2}\right) = \left(\frac{z}{2}\right)^2 - 6$$
    $$= \frac{z^2}{4} - 6$$

The function can be evaluated for a constant, such as $f(2)$; for a variable, such as $f(a)$; or for another expression, such as $f(b + 1)$.

FUNCTIONS WITH MORE THAN ONE VARIABLE

A function can be defined for multiple variables: for example, $f(a, b) = a + 2b$.

To evaluate $f(2, 3)$, the value for a is 2, and the value for b is 3.

$$f(2, 3) = 2 + 2(3) = 8$$

To evaluate $f(3, 2)$, the value for a is 3 and the value for b is 2.

$$f(3, 2) = 3 + 2(2) = 7$$

Defining a function such as $f(a, b, c) = (a + b)/c$, with the variables identified as $a = 2$, $b = 6$, and $c = 4$, then

$$f(2, 6, 4) = \frac{2 + 6}{4}$$

$$= \frac{8}{4} = 2$$

When the variables are $a = 4$, $b = 6$, and $c = 2$, then

$$f(4, 6, 2) = \frac{4 + 6}{2}$$

$$= \frac{10}{2} = 5$$

When the variables are $a = 2$, $b = 4$, and $c = 6$, then

$$f(2, 4, 6) = \frac{2 + 4}{6}$$

$$= \frac{6}{6} = 1$$

The function can also be evaluated for other expressions:

$$f(x + 1, 5x + 3, 2) = \frac{(x + 1) + (5x + 3)}{2}$$

$$= \frac{6x + 4}{2}$$

$$= 3x + 2$$

EXERCISES

1. Which of these relations are functions?
   (a) $\{(1, 4), (2, 4), (3, 8)\}$
   (b) $\{(2, 3), (3, 2), (4, 5), (5, 4)\}$
   (c) $\{(1, 1), (2, 2), (3, 3), (4, 4)\}$
   (d) $\{(2, 4), (3, 8), (4, 16), (5, 32)\}$
   (e) $\{(0, 1), (1, 1), (2, 2)\}$

2. Identify the domain and range for

$$\{(0, 0), (1, 2), (2, 4), (3, 6), (4, 8)\}$$

3. Given the function $f(x) = 7x - 4$ and the range $\{-2, 0, 3, 5, 6\}$, find the domain.

4. Given the function $f(x) = 11 - 3x$, find:
   (a) $f(-2)$ (b) $f(4)$
   (c) $f(a)$ (d) $f(c + d)$
   (e) $f(4g)$
   (g) $f(.5)$ (f) $f\left(\dfrac{b}{2}\right)$
   (i) $f(2a - b)$
                            (h) $f(a + 7)$

                            (j) $f\left(\dfrac{a}{6} - 3\right)$

5. Given the function $f(a, b) = 3 + 5a - 2b$, find:
   (a) $f(3, 4)$ (b) $f(4, 3)$
   (c) $f(1, a)$ (d) $f(b, 3)$
   (e) $f(-2, 4)$ (f) $f(c + 2, d - 1)$
   (g) $f(.4, .25)$

OPERATIONS WITH FUNCTIONS

Functions can be written as $f(x)$, $g(x)$, $h(x)$, and so on. The function variable does not always have to be identified as f. Operations with functions are the same as operations with numbers:

Addition: $f(x) + g(x)$
Subtraction: $f(x) - g(x)$
Multiplication: $f(x) \times g(x)$

Division: $\dfrac{f(x)}{g(x)}$

Examine the following two functions:

$$f(x) = 3x - 4 \qquad g(x) = 5 - 2x$$

To evaluate the addition of the two functions:

$$f(x) + g(x) = (3x - 4) + (5 - 2x)$$
$$= 3x - 4 + 5 - 2x$$
$$= x + 1$$

To evaluate the subtraction of the two functions:

$$f(x) - g(x) = (3x - 4) - (5 - 2x)$$
$$= 3x - 4 - 5 + 2x$$
$$= 5x - 9$$

To evaluate the multiplication of the two functions:

$$f(x) \times g(x) = (3x - 4) \times (5 - 2x)$$
$$= 3x(5) - 3x(2x) - 4(5) - 4(-2x)$$
$$= 15x - 6x^2 - 20 + 8x$$
$$= -6x^2 + 23x - 20$$

To evaluate the division of the two functions:

$$\frac{f(x)}{g(x)} = \frac{3x - 4}{5 - 2x}$$

These functions can also be evaluated for constants, variables, and expressions.

Using a constant gives us

$$f(8) \times g(2) = (3(8) - 4) \times (5 - 2(2))$$
$$= (24 - 4) \times (5 - 4)$$
$$= 20 \times 1$$
$$= 20$$

Using an expression gives us

$$f(a + b) + g(a - b) = 3(a + b) - 4 + 5 - 2(a - b)$$
$$= 3a + 3b - 4 + 5 - 2a + 2b$$
$$= a + 5b + 1$$

COMPOSITE FUNCTIONS

Composite functions are written as $f(g(x))$. Here the function g becomes the domain of the function f. In the case of $f(x) = 5x + 2$ and $g(x) = 8 - 3x$, then

$$f(g(x)) = 5(8 - 3x) + 2$$
$$= 40 - 15x + 2$$
$$= 42 - 15x$$

and

$$f(g(3)) = 42 - 15(3)$$
$$= 42 - 45$$
$$= -3$$

Similarly, the composite function can be $g(f(x))$. Evaluating this composite function would be

$$g(f(x)) = 8 - 3(5x + 2)$$
$$= 8 - 15x - 6$$
$$= 2 - 15x$$
$$g(f(3)) = 2 - 15(3)$$
$$= 2 - 45$$
$$= -43$$

FUNCTIONS IN PROGRAMMING

The functions that are available to the programmer are specified in the language of the program. These are stated so that the computer recognizes them. Although some functions will vary from one computer language and computer system to another, there are certain functions that are common to most languages.

In mathematical terms, a function can be specified as $f(x) = \sqrt{x}$, so $f(9) = 3$. But in computer terms, a function is usually specified as SQRT (X), so SQRT(9) = 3.

Another function is $f(x) = |x|$, which is the absolute value function. In computer notation, it is specified as ABS(X). Therefore, where $f(-3) = 3$, the absolute value function appears as ABS(-3) = 3. The functions in the computer define a unique value for the domain element given.

A function can be more than just the evaluation of an equation. It can also be a conditional. For example, the function

$$SGN(X) = \begin{cases} -1 & \text{if } X < 0 \\ 0 & \text{if } X = 0 \\ 1 & \text{if } X > 0 \end{cases} \qquad \text{Thus:} \begin{cases} SGN(2) = 1 \\ SGN(-8) = -1 \\ SGN(0) = 0 \end{cases}$$

EXERCISES

1. Given $f(x) = 3x + 7$ and $g(x) = x^2 - 2x + 1$, find:
   (a) $f(-3)$ (b) $g(-3)$
   (c) $f(x) + g(x)$ (d) $f(x) - g(x)$
   (e) $f(5) - g(3)$ (f) $f(-1) + g(2)$
   (g) $g(x) - f(x)$ (h) $f(a + 2) - g(a)$

2. Given $f(x) = 7x + 8$ and $g(x) = 3x - 2$, find:
   (a) $f(x) \times g(x)$ (b) $(f(x) + g(x))/2$
   (c) $f(-2) \times g(3)$
   (e) $(f(x) - g(x)) \times g(x)$ (d) $\dfrac{f(2)}{g(4)}$
   (g) $f(a - 1) \times g(a + 1)$
                                 (f) $\dfrac{g(4)}{f(2)}$

3. Given $f(x) = x + 2$ and $g(x) = x^2 - 3x + 7$, find:
   (a) $f(g(x))$ (b) $g(f(x))$
   (c) $f(g(2))$ (d) $g(f(2))$
   (e) $f(f(x))$ (f) $g(g(x))$

4. Using the following table, evaluate the functions shown after the table.

	Algebraic	Computer
	$f(x) = \sqrt{x}$	SQRT (X)
	$f(x) = \lvert x \rvert$	ABS (X)
$f(x) =$	$\begin{cases} -1 & \text{if } x < 0 \\ 0 & \text{if } x = 0 \\ 1 & \text{if } x > 0 \end{cases}$	SGN (X)
	$f(x) = e^x$	EXP (X)
	$(e \approx 2.71828)$	

(a) SQRT(25)   (b) ABS(7 - 9)
(c) SGN(8 - 3)   (d) EXP(1)
(e) SQRT(ABS(16 - 25))   (f) SGN(ABS(-3))
(g) EXP(0)   (h) EXP(SGN(11 - 4))
(i) SGN(EXP(-12))   (j) SQRT(EXP (2))

5. Research a computer language you are familiar with and give a list of functions that are intrinsic to the language.

S U M M A R Y _____

Relations are sets of ordered pairs. Functions in mathematics are relations such that for each value of x, there is only one value for $f(x)$, or y. The values for x are the domain and the values for $f(x)$ are the range.

The domain can be a constant, a variable, or an expression. These are used to calculate the range of the function.

Functions can have more than one variable. The variables specified are the variables in the function. For example, $f(a, x, b) = ax + b$ is a function with three variables.

The operations performed on functions are:

Addition:	$f(x) + g(x)$
Subtraction:	$f(x) - g(x)$
Multiplication:	$f(x) \times g(x)$
Division:	$\dfrac{f(x)}{g(x)}$

The composite function is a function used as a variable for another function, such as $f(g(x))$.

The computer programmer has two types of functions available in programming:

1. Functions that are intrinsic to the language. They operate as functions but are in a different format.

2. Functions that are defined by the programmer. They are similar to mathematical functions in that—given a variable or variables—the function defines a unique value.

C H A P T E R   R E V I E W _____

1. A set of ordered pairs is a _____ .

2. A _____ is a relation such that for a value of x, there exists a unique value of y.

3. Given $\{(2, 4), (2, 7), (5, 2), (4, 2)\}$. The relation is _____ a function because _____ .

4. The _____ of a function is the set of x values.

5. The _____ of a function is the set of y values.

6. Given the function $f(x) = 5x + 3$, then $f(-2) =$ ＿＿＿ and $f(a + 3) =$ ＿＿＿ .

7. If $f(a, c) = 3a + 2c$, to evaluate $f(2, 3)$ means that $a =$ ＿＿＿ and $c =$ ＿＿＿ .

8. Function operations are ＿＿＿＿＿＿＿＿ , ＿＿＿＿＿＿＿＿ , ＿＿＿＿＿＿＿＿ , and ＿＿＿＿＿＿＿＿ .

9. $f(g(x))$ is a ＿＿＿＿＿＿＿ function.

10. $f(x) = \sqrt{x}$ as a computer function might be ＿＿＿＿＿＿ .

Sequence
and Subscripted
Variables

chapter 15

A *sequence* is a function whose domain is a set of positive integers. The sequence values are the range of the function. A sequence is usually referred to as an *array* and as such is generated by a loop. The loop values are restricted to positive integers and as such identify the subscript of the array.

Arrays can be used in a computer to store large amounts of data. The data stored in an array can be manipulated, which can include sorting and updating.

With an array, the ordered pair (2, 5) can be written as S(2) = 5. For domain value of 2, the function S has a range value of 5. As a function, no two x values can be the same for different y values in an array. For example, it would be impossible to have S(2) = 5 and S(2) = 6.

SEQUENCE DEFINITION

A sequence is a function whose domain is a set of positive integers. Mathematically, the sequence is written as

$$S_1 = 2$$
$$S_2 = 4$$
$$S_3 = 6$$
$$S_4 = 8$$
$$S_5 = 10$$

The domain values $\{1, 2, 3, 4, 5\}$ are called the *subscripts* of the sequence. The function values $\{2, 4, 6, 8, 10\}$ are the range. In this chapter the function notations more suited to programming will be used:

$$S(1) = 2$$
$$S(2) = 4$$
$$S(3) = 6$$
$$S(4) = 8$$
$$S(5) = 10$$

A sequence can be defined in two ways:

1. *N-term:* In the given sequence $S(1)$ is two times the subscript 1. $S(2)$ is two times the subscript 2, and similarly for all the terms in the sequence. So the N-term definition of the sequence is

$$S(N) = 2 * N$$

2. *Recursive:* With the recursive definition, one can observe how each successive term relates to the preceding term. For example, $S(1) = 2$; $S(2) = S(1) + 2$, or $2 + 2 = 4$; $S(3)$ is $S(2) + 2$, or $4 + 2 = 6$; and so on. Therefore, the recursive definition is

$$S(1) = 2$$
$$S(N) = S(N - 1) + 2$$

In general, sequences can be defined by using either method.

The N-term definition has an advantage over the recursive definition because $S(100)$ can be found in one step, whereas in the recursive definition, $S(100)$ means finding $S(1)$ through $S(99)$.

EXAMPLE 1

Find the terms of a sequence with domain $\{1, 2, 3, 4, 5, \ldots, 10\}$ for $S(N) = 3 * N + 1$.

Solution   The terms of the sequence are

S(1) = 4	S(6) = 19
S(2) = 7	S(7) = 22
S(3) = 10	S(8) = 25
S(4) = 13	S(9) = 28
S(5) = 16	S(10) = 31

EXAMPLE 2

Using the sequence S(N) = 3 * N + 1, find the recursive definition.

Solution:

$$S(1) + 3 \text{ gives } S(2)$$
$$S(2) + 3 \text{ gives } S(3)$$
$$S(3) + 3 \text{ gives } S(4)$$

The recursive definition is

$$S(1) = 4$$
$$S(N) = S(N - 1) + 3$$

EXAMPLE 3

Given the recursive definition, S(1) = 5, S(N) = S(N - 1) - 2; find the first 10 terms of the sequence.

Solution:

S(1) = 5	S(6) = -3 - 2 = -5
S(2) = 5 - 2 = 3	S(7) = -5 - 2 = -7
S(3) = 3 - 2 = 1	S(8) = -7 - 2 = -9
S(4) = 1 - 2 = -1	S(9) = -9 - 2 = -11
S(5) = -1 - 2 = -3	S(10) = -11 - 2 = -13

EXAMPLE 4

Find the N-term definition of the sequence S(1) = 5, S(N) = S(N - 1) - 2.

Solution   One method is to think in terms of multiplying and adding. That is, in general, the N-term definition will include some multiplication and addition: S(N) = A * N + B.

Using S(1) = 5 for N = 1 gives the equation

$$A * 1 + B = 5$$
$$A + B = 5$$
$$A = 5 - B$$

Selecting any of the other terms in the sequence $S(4) = -1$ for $N = 4$ gives the equation $A * 4 + B = -1$. Substituting $5 - B$ for A gives

$$(5 - B) * 4 + B = -1$$
$$20 - 4B + B = -1$$
$$20 - 3B = -1$$
$$-3B = -21$$
$$B = 7$$
$$A = 5 - B = 5 - 7$$
$$A = -2$$

The N-term definition is

$$S(N) = -2 * N + 7$$

To check: Pick an N value less than 10 and use the terms generated by Example 3:

$$N = 7$$
$$S(7) = -2 * 7 + 7$$
$$= -14 + 7 = -7$$

EXAMPLE 5

Find the recursive definition for the following sequence:

$$\{-1, 1, 5, 13, 29\}$$

Solution   To find the recursive definition, find the difference between successive terms:

$$S(2) - S(1) = 1 - (-1) = 2 \quad \rightarrow \quad S(2) = S(1) + 2$$
$$S(3) - S(2) = 5 - 1 = 4 \quad \quad \rightarrow \quad S(3) = S(2) + 4$$
$$S(4) - S(3) = 13 - 5 = 8 \quad \quad \rightarrow \quad S(4) = S(3) + 8$$
$$S(5) - S(4) = 29 - 13 = 16 \rightarrow \quad S(5) = S(4) + 16$$

2, 4, 8, and 16 represent powers of 2.

$$2 = 2 \qquad \text{So, for N = 2, the power is 1}$$
$$2^2 = 4 \qquad \text{for N = 3, the power is 2}$$
$$2^3 = 8 \qquad \text{for N = 4, the power is 3}$$
$$2^4 = 16 \qquad \text{for N = 5, the power is 4}$$

In general for N, the power is N - 1. The recursive definition is

$$S(1) = -1$$
$$S(N) = S(N - 1) + 2 ** (N - 1)$$

EXERCISES

1. Fill in the missing terms.
   (a) 25, 20, 15, 10, ____ , ____ , ____ .
   (b) 5, 9, 13, 17, ____ , ____ , ____ .
   (c) 1, -2, 4, -8, 16, ____ , ____ , ____ .
   (d) 1, 3, 7, 15, 31, ____ , ____ , ____ .
   (e) 9, 7, 3, -3, -11, ____ , ____ , ____ .

2. Find the first five terms of each sequence given the N-term definition.
   (a) $S(N) = N * 3$ (b) $S(N) = 5 * N - 4$
   (c) $S(N) = N * (N + 1)$ (d) $S(N) = 15 - 3 * N$
   (e) $S(N) = N ** 2 - N$ (f) $S(N) = (N - 1)(N + 1)$
   (g) $S(N) = 2 ** N$ (h) $S(N) = 1/N$
   (i) $S(N) = N/(N + 1)$ (j) $S(N) = (-1) ** N + 1$

3. Find the first five terms of each sequence, given the recursive definition.
   (a) $S(1) = 3$ (b) $S(1) = 1$
       $S(N) = S(N - 1) + 4$ $S(N) = 2 * S(N - 1)$
   (c) $S(1) = 8$ (d) $S(1) = 2$
       $S(N) = 20 - S(N - 1)$ $S(N) = S(N - 1) ** 2$
   (e) $S(1) = -5$ (f) $S(1) = 100$
       $S(N) = 100 - 4 * S(N - 1)$ $S(N) = S(N - 1)/2$
   (g) $S(1) = 100$ (h) $S(1) = 1$
       $S(N) = S(N - 1)/5$ $S(2) = 1$
   (i) $S(1) = 1$ $S(N) = S(N - 1) + S(N - 2)$
       $S(2) = 2$
       $S(N) = S(N - 1) * S(N - 2)$

4. Find the N-term definition for each sequence.
   (a) 3, 6, 9, 12, 15   (b) 1, 4, 9, 16, 25
   (c) 1, 3, 5, 7, 9   (d) 3, 5, 7, 9, 11
   (e) 4, 9, 14, 19, 24   (f) 7, 10, 13, 16, 19
   (g) 10, 8, 6, 4, 2   (h) 20, 25, 30, 25, 40

5. Find the recursive definition for each sequence.
   (a) 5, 10, 15, 20, 25   (b) 2, 5, 8, 11, 14
   (c) 1, 2, 4, 8, 16   (d) 2, 6, 18, 54, 162
   (e) 3, 8, 18, 38, 78   (f) 4, 5, 7, 11, 19
   (g) 1, -1, 1, -1, 1   (h) 3, 9, 81, 6561, 43046721

6. Find the N-term and recursive definitions for the set of odd numbers

$$1, 3, 5, 7, \ldots$$

7. Find the missing terms and find the N-term and recursive definitions for the sequence

$$1, 4, \text{\_\_\_\_}, 10, 13, \text{\_\_\_\_}, \text{\_\_\_\_}$$

8. Given the sequence 4, 9, 14, 19, 24, 29, 34, . . . , 104, find the three parts of a loop to generate the numbers.

ARITHMETIC SEQUENCES

In an *arithmetic* sequence the terms of the sequence have a *common difference*. If $S(1) = 2$,

$$S(2) = 7$$
$$S(3) = 12$$
$$S(4) = 17$$
$$S(5) = 22$$
$$S(6) = 27$$
$$S(7) = 32$$

The difference between any two succeeding terms is constant.

$$S(2) - S(1) = 7 - 2 = 5$$
$$S(3) - S(2) = 12 - 7 = 5$$
$$S(4) - S(3) = 17 - 12 = 5$$
$$S(5) - S(4) = 22 - 17 = 5$$

Therefore, the sequence is an arithmetic sequence.

To find the terms of an arithmetic sequence, certain constants must be known: for example, $S(1)$, the first term of the sequence; and D, the common difference. Knowing these constants, the formula for any term in the arithmetic sequence is

$$S(N) = S(1) + (N - 1) * D$$

Continuing with this example, the given sequence has $S(1) = 2$ and $D = 5$. To find the $S(25)$ term, use the formula

$$S(25) = 2 + (25 - 1) * 5$$
$$= 2 + 24 * 5$$
$$= 2 + 120$$
$$= 122$$

The formula is the standard N-term definition for an arithmetic sequence.
The recursive definition is

$$S(1) = 2$$
$$S(N) = S(N - 1) + 5$$

An arithmetic sequence can be generated by a computer using either the domain or the range. For example, use a loop to generate the terms using the domain for the sequence

$2, 7, 12, 17, 22, 27, 32, \ldots$ up to and including the 100th term

Here the N-term definition should be used:

$$S(1) = 2$$
$$S(N) = S(1) + (N - 1) * D$$

The algorithm is:

Step 1:   Input $S(1) = 2$ and $D = 5$
          Initialize loop at $N = 1$
Step 2: $S(N) = S(1) + (N - 1) * D$
Step 3:   Increment N by 1
Step 4:   Test: Is $N > 100$?
                  Yes: Go to step 5
                  No: Go to step 2
Step 5:   Stop

The flowchart is

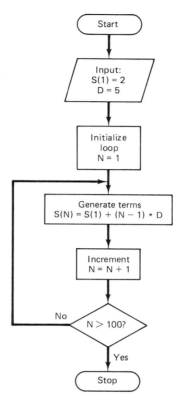

A loop can be used to generate the terms of an array if the following is given:

$$A(1) = 2$$

$$D = 5$$

$$A(N) = 202$$

In this case, the algorithm is based on the following N-term definition:

Step 1:   Set N = 1
Step 2:   Initialize loop: I = 2
Step 3:   S(N) = I
Step 4:   N = N + 1
Step 5:   Increment I by 5: I = I + 5

Step 6:   Test: Is I > 202?
Yes: Go to step 7
No: Go to step 3

Step 7:   Stop

For the domain of N = 1, 2, 3, . . . , it is necessary to set a different variable for the range.

The flowchart is

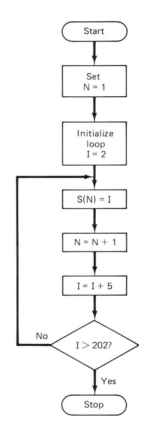

EXAMPLE

Given the terms S(3) = 5 and S(9) = 8, find the S(100) term of an arithmetic sequence.

Solution:

1. Find the S(1) and the common difference in one step. Using the N-term definition, we have

$$S(1) + (3 - 1) * D = 5$$

$$S(1) + (9 - 1) * D = 8$$

$$S(1) + 2 * D = 5 \quad \text{or} \quad S(1) = 5 - 2 * D$$

$$S(1) + 8 * D = 8 \quad \text{substituting } S(1)$$

$$5 - 2 * D + 8 * D = 8$$

$$5 + 6 * D = 8$$

$$6 * D = 3$$

The common difference is D = 1/2 or .5.

$$S(1) = 5 - 2 * 1/2$$

$$= 5 - 1$$

The S(1) term is: = 4

2. To find the S(100) term:

$$S(100) = S(1) + (100 - 1) * 1/2$$

$$= 4 + 99 * 1/2$$

$$= 4 + 49.5$$

The S(100) term is: = 53.5

EXERCISES

1. Find the first five terms for each arithmetic sequence. The first term is S(1) and D is the common difference.

    (a) S(1) = 8   D = 2   (b) S(1) = 5   D = 3
    (c) S(1) = -3   D = 4   (d) S(1) = 225   D = 5
    (e) S(1) = -1/2   D = 2   (f) S(1) = 1/2   D = 1/4
    (g) S(1) = 2/3   D = 1/3   (h) S(1) = 100   D = 8
    (i) S(1) = 2   D = -4   (j) S(1) = 24   D = -6

    Given the following arithmetic sequences:

    A. 1, 4.5, 8, 11.5, 15, 18.5, 22, . . .

    B. 24, 21, 18, 15, 12, 9, 6, 3, 0, . . .

    C. -8, -5.52, -3.04, -0.56, 1.92, . . .

2. Find the N-term definition of each.
3. Find the recursive definition of each.

4. Find the S(50) term for each.
5. Given S(1) = -10 and D = .25, write an algorithm to generate the first 50 terms of the sequence.
6. Flowchart the algorithm for Exercise 5.
7. Given S(1) = 100 and D = -2, write an algorithm to generate the first N terms up to S(N) = -96.
8. Flowchart the algorithm for Exercise 7.
9. Given S(7) = 27 and S(10) = 42, find the S(100) term.
10. Find the recursive definition of any arithmetic sequence given the S(1) and the D terms.

GEOMETRIC SEQUENCES

A geometric sequence is one in which the terms have a *common ratio.* If S(1) = 5,

$$S(2) = 10$$
$$S(3) = 20$$
$$S(4) = 40$$
$$S(5) = 80$$
$$S(6) = 160$$

The ratio between any two succeeding terms is constant.

$$S(2)/S(1) = 10/5 = 2$$
$$S(3)/S(2) = 20/10 = 2$$
$$S(4)/S(3) = 40/20 = 2$$
$$S(5)/S(4) = 80/40 = 2$$
$$S(6)/S(5) = 160/80 = 2$$

The common ratio is 2 and the first term is 5.

Given the S(1) and the common ratio R, the N-term definition of a geometric sequence is

$$S(N) = S(1) * R ** (N - 1)$$

EXAMPLE 1

Given S(1) = 6 and R = 3, find the first five terms of the sequence.

Solution:

$$S(1) = 6 * 3 ** 0 = 6$$
$$S(2) = 6 * 3 ** 1 = 18$$
$$S(3) = 6 * 3 ** 2 = 54$$
$$S(4) = 6 * 3 ** 3 = 162$$
$$S(5) = 6 * 3 ** 4 = 486$$

Using the values given earlier, find the recursive definition of the geometric sequence.

$$S(1) = 6$$
$$S(2) = 18$$
$$S(3) = 54$$
$$S(4) = 162$$
$$S(5) = 486$$

The successive terms are multiples of the previous term:

$$6 * 3 = 18$$
$$18 * 3 = 54$$
$$54 * 3 = 162$$

The recursive definition of the sequence is

$$S(1) = 6$$
$$S(N) = S(N - 1) * 3$$

EXAMPLE 2

Given S(4) and S(10), find the S(6) term of a geometric sequence.

$$S(4) = 13.5 \qquad S(10) = 9841.5$$

Solution   Using the N-term definition, find the S(1) and the R terms.

$$S(4) = 13.5 = S(1) * R ** 3$$
$$S(10) = 9841.5 = S(1) * R ** 9$$

Dividing:

$$S(10)/S(4) = 729 = R ** 6$$

Find the sixth root of 729:

$$R = 729 ** (1/6)$$
$$= 3$$

To find S(1), substitute R = 3 in the S(4) equation:

$$13.5 = S(1) * R ** 3$$
$$= S(1) * 3 ** 3$$
$$= S(1) * 27$$
$$S(1) = 13.5/27$$
$$= .5$$

To generalize, find the algorithm to determine the S(1) and the R, given the S(A) and the S(B) terms.

Input: A, B, S(A), and S(B)
Process: Step 1:   Divide: Q = S(B)/S(A)
          Step 2:   Subtract: P = B − A
          Step 3:   Raise to a power: R = Q ** (1/P)
          Step 4:   Divide: S(1) = S(A)/R ** (A − 1)
Output: S(1) and R

EXAMPLE 3

Determine N given the following:

$$S(1) = 2$$
$$R = 3$$
$$S(N) = 39366$$

Solution   Using the N-term definition gives us

$$39366 = 2 * 3 ** (N − 1)$$

Then we divide by 2:

$$19683 = 3 ** (N − 1)$$

To find N, use the LOG function. (The LOG function is found in most computer languages.)

Take the LOG of both sides:

$$LOG(19683) = LOG(3 ** (N - 1))$$

$$LOG(19683) = (N - 1) * LOG(3)$$

$$LOG (19683)/LOG(3) = N - 1$$

$$LOG(19683)/LOG(3) + 1 = N$$

$$N = 10$$

The S(10) term is 39366.

EXERCISES

1. Determine which sequences are geometric.
   (a) 8, 4, 2, 1, 1/2, 1/4, . . .
   (b) 100, 20, 4, 0.8, 0.16, 0.032, . . .
   (c) 5, 10, 15, 20, 25, 30, . . .
   (d) .5, 1, 2, 4, 8, 16, 32, . . .
   (e) 10, 1, 0.1, 0.01, 0.001, 0.0001, . . .

2. Find the first five terms of the geometric sequences given the S(1) term and the R ratio.

(a) S(1) = 75	R = 1/3	(b) S(1) = 5	R = 2
(c) S(1) = 16	R = 1/2	(d) S(1) = 100	R = 1/2
(e) S(1) = 100	R = 2	(f) S(1) = 1000	R = .1
(g) S(1) = .0001	R = 10	(h) S(1) = 2	R = .25
(i) S(1) = 666	R = 1/3	(j) S(1) = 25	R = .2

3. Given the geometric sequence: 6, 1.8, .54, .162, .0486, . . .
   (a) Find the common ratio.   (b) Find the S(8) term.
   (c) Find the recursive definition.   (d) Write an algorithm to generate
   (e) Flowchart the algorithm.   the first 100 terms of the se-
   (f) Find the N for which S(N) =   quence.
        .000118098.

4. Given S(5) = .08 and S(9) = 1.28, find the S(20) of the geometric sequence.

5. Given S(1) = 81, R = 1/3, and S(N) = .0001524, find the value for N.

THE SEQUENCE AS AN ARRAY

A sequence is a function which translates into an array for the programmer. The array is a list of numbers identified by an array name and a subscript. The subscript must be an integer. In the ordered pair formed by (subscript, array element) no subscripts can be the same for two different array elements.

The elements of an array can be assigned by a loop; for example, the flowchart to generate the first 100 terms for any array named "NUM(N)" would be

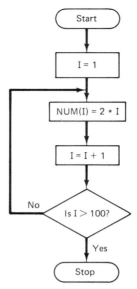

The flowchart assigns the numbers to the array "NUM(N)":

$$NUM (1) = 2$$

$$NUM (2) = 4$$

$$NUM (3) = 6$$

$$NUM (4) = 8$$

$$NUM (5) = 10$$

$$NUM (6) = 12$$

$$\vdots$$

$$NUM (100) = 200$$

The array is used to store lists of numbers so that the numbers can be manipulated within a program.

EXAMPLE

A retail store stocked 100 of item 402. Each monthly order for that item was increased by 5%. How many of the items were stocked in the twelfth month? What was the average order for the 12 months?

Solution   If the order increases by 5% each month, projections for the first and following months will be as follows:

$$ORDER (1) = 100$$

$$ORDER (2) = 105$$

$$ORDER\ (3) = 110.25$$

$$\vdots$$

$$ORDER\ (N) = ORDER\ (N - 1) * .05 + ORDER(N - 1)$$

To find the average order for 12 months, add the elements of the **array** and divide by 12. This means that a summation must be used.

The flowchart is as follows:

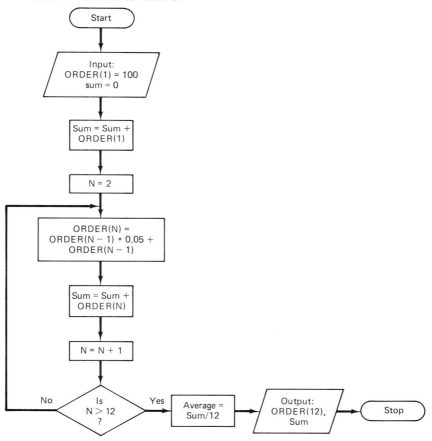

Look at the terms of the array:

$$ORDER\ (1) = 100$$

$$ORDER\ (2) = 105$$

$$ORDER\ (3) = 110.25$$

$$ORDER\ (4) = 115.7625$$

$$\vdots$$

There is no common difference, but there is a common ratio.

$$100/105 = .952381$$

$$105/110.25 = .952381$$

$$110.25/115.7625 = .952381$$

Therefore, this is a geometric sequence. The geometric sequence is

$$S(1) = 100$$

If $S(2) = 105$, then

$$105 = 100 * R$$

$$1.05 = R$$

The N-term definition is

$$S(N) = 100 * (1.05) ** (N - 1)$$

EXERCISES

1. Given the following flowcharts, determine the first and last terms of each array.

(a)

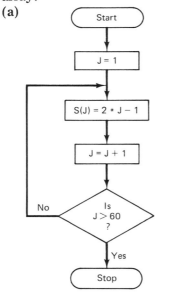

(b)

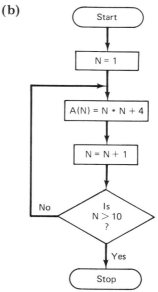

(c)

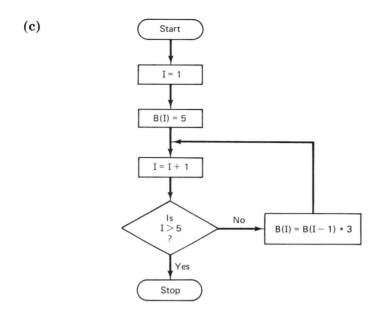

2. Given the following flowchart, fill in the list of values in array PROD(N).

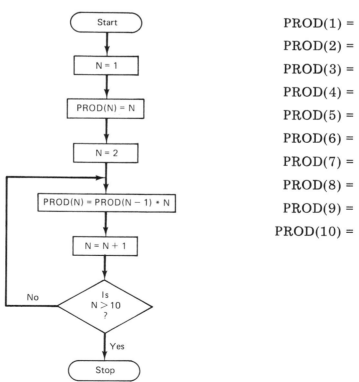

PROD(1) =

PROD(2) =

PROD(3) =

PROD(4) =

PROD(5) =

PROD(6) =

PROD(7) =

PROD(8) =

PROD(9) =

PROD(10) =

3. Draw a flowchart to generate the sequence

$$5, 5.25, 5.50, 5.75, 6, \ldots$$

4. Draw a flowchart to solve the following problem. It takes 10 seconds for a car to roll down an incline. In the first second the car rolls 7 feet. Each second thereafter the car rolls 10 feet. How long is the incline?

5. The Piranha Loan Company will loan you $100. You are given 10 months to repay the loan. The terms are: $1 the first month, $2 the second month, $4 the third month, $8 the fourth month, and so on, for the remaining months. Draw a flowchart to determine how much is paid each month and the total paid back on the loan.

6. Draw a flowchart to determine how many boxes are needed for a display stacked, such that

$$Row\ (1) = 15$$
$$Row\ (2) = 14$$
$$Row\ (3) = 13$$
$$\vdots$$
$$Row\ (N) = 1$$

SUMMARY _____

The sequence is a function in which the domain is the set of positive integers. The sequence can be defined as $S(N) = 3 * N - 5$, which is the N-term definition. Using the N-term definition, it is possible to find any element—in sequence—independent of the others.

The recursive definition of a sequence lists the first term and the succeeding terms in relation to the preceding term. For $S(N) = 3 * N - 5$, the recursive definition is

$$S(1) = -2$$
$$S(N) = S(N - 1) + 3$$

Using the recursive definition, it is necessary to find the $(N - 1)$ terms before the N term.

The arithmetic and geometric sequences are special. The arithmetic sequence is defined as

$$S(N) = S(1) + (N - 1) * D$$

where the terms have a common difference. The geometric sequence is defined as

$$S(N) = S(1) * R ** (N - 1)$$

where the terms have a common ratio.

     To the programmer, the sequence is an array. It is a list in which each element is identified by its subscript. The subscripts must be integers and each subscript must identify a unique element. Arrays are used to store data so that it will be available for processing.

CHAPTER REVIEW _____

1. A sequence is a function whose domain is the set of _____ .
2. The range is the set of _____ .
3. Given a sequence $\{1, 3, 5, 7, 9, \ldots\}$, S(1) is _____ and S(N) is _____ .
4. The recursive definition of the sequence $\{1, 3, 5, 7, 9, \ldots\}$ is

$$S(1) \ \_\_\_\_$$

$$S(N) \ _____$$

5. A sequence is an arithmetic sequence if the terms have a _____

     _____ .

6. The N-term definition of an arithmetic sequence is S(N) = _____ .
7. The recursive definition of an arithmetic sequence is S(N) = _____

     _____ .

8. A sequence is a geometric sequence if the terms have a _____ .
9. The N-term definition of a geometric sequence is S(N) = _____

     _____ .

10. To the programmer, the sequence is known as an _____ .

Matrices

The sequence, which is a list of numbers, can be considered as an array. The array has one subscript, which means that it can be defined as a one-dimensional array. Matrices are numbers in a grid. These numbers are stored in a row as well as a column. The programmer redefines the matrix and calls it an array. But because the matrix has two dimensions, it is a two-dimensional array. Elements stored in this array have two subscripts: one for row, and the second for column.

The matrix as an array is used to store data. The subscripts will represent *two identifications* for each data element. For example, the numbers of item 2, stored in warehouse 3, would be shown in an array as ITEM (2, 3).

The programmer can use certain matrix operations in the form of function definitions. In this way, data can be manipulated within the array. (This is also referred to as *table handling.*)

THE MATRIX

A *matrix* is a grid representing numbers stored in rows and columns. Mathematically, matrix **A**, consisting of two rows and three columns, is written as

$$A = \begin{bmatrix} 5 & 7 & 9 \\ 12 & 15 & 24 \end{bmatrix}$$

The elements are each identified by subscripts for row and column.

The matrix **A** is dimensioned as DIMENSION A(2, 3). The number of rows is shown as the first subscript, and the number of columns is shown as the second.

The elements are each identified by their location and matrix:

$A(1, 1) = 5$ (first row, first column)

$A(2, 1) = 12$ (second row, first column)

$A(2, 3) = 24$ (second row, third column)

Two matrices are equal if:

1. The dimensions are the same.
2. Each element of one is equal to each element of the other.

EXAMPLE

$$A = \begin{bmatrix} 1 & 3 \\ 5 & 7 \\ 9 & 11 \end{bmatrix} \qquad B = \begin{bmatrix} 1 & 3 \\ 5 & 7 \\ 9 & 11 \end{bmatrix}$$

The dimensions of **A** are: 3 rows $\times$ 2 columns.

The dimensions of **B** are: 3 rows $\times$ 2 columns.

Therefore, both dimensions are the same. The elements of **A** and **B** are also equal, since

$$A(1, 1) = B(1, 1)$$
$$A(1, 2) = B(1, 2)$$
$$A(2, 1) = B(2, 1)$$
$$A(2, 2) = B(2, 2)$$
$$\vdots \qquad \vdots$$

Therefore, matrix **A** = matrix **B**.

MATRIX OPERATIONS OF ADDITION AND SUBTRACTION

Two matrices can be added if the dimensions of both are the same. The addition of two matrices is defined as the sum of the individual elements.

To add matrices **A** and **B**, and store the sum in matrix **D**, the formula is

$$R = \text{number of rows}$$

$$C = \text{number of columns}$$

$$D(I, J) = A(I, J) + B(I, J)$$

where

$$I = \{1, 2, \ldots\ldots, R\}$$

$$J = \{1, 2, \ldots\ldots, C\}$$

The algorithm is:

Step 1:   Set row counter to 1:

$$I = 1$$

Step 2:   Set column counter to 1:

$$J = 1$$

Step 3:   Add elements:

$$D(I, J) = A(I, J) + B(I, J)$$

Step 4:   Increment column counter:

$$J = J + 1$$

Step 5:   Test for end of column:
Is $J > C$?
No: Go to step 3
Yes: Go to step 6

Step 6:   Increment row counter:

$$I = I + 1$$

Step 7: Test for end of row:
Is $I > R$?
No: Go to step 2
Yes: Stop

EXAMPLE 1

Add matrices **A** and **B**.

$$C = A + B$$

$$A = \begin{bmatrix} 4 & 3 \\ 7 & 5 \\ 6 & 8 \end{bmatrix} \qquad B = \begin{bmatrix} 10 & 12 \\ 27 & 32 \\ 45 & 66 \end{bmatrix}$$

Solution Using the algorithm:

$$C = \begin{bmatrix} 4 + 10 & 3 + 12 \\ 7 + 27 & 5 + 32 \\ 6 + 45 & 8 + 66 \end{bmatrix}$$

$$= \begin{bmatrix} 14 & 15 \\ 34 & 37 \\ 51 & 74 \end{bmatrix}$$

EXAMPLE 2

Subtract:

$$D = B - A$$

Solution:

$$D = \begin{bmatrix} 10 - 4 & 12 - 3 \\ 27 - 7 & 32 - 5 \\ 45 - 6 & 66 - 8 \end{bmatrix}$$

By definition, multiplication is repeated addition. So when multiplying a matrix by a number, use repeated addition.

$$2 * \text{matrix } A = A + A$$

$$3 * \text{matrix } B = B + B + B$$

Therefore, the product of a number times a matrix is each element of the matrix multiplied by the number.

$$2 * \begin{bmatrix} 1 & 3 \\ 5 & 7 \\ 9 & 11 \end{bmatrix} = \begin{bmatrix} 2*1 & 2*3 \\ 2*5 & 2*7 \\ 2*9 & 2*11 \end{bmatrix}$$

$$= \begin{bmatrix} 2 & 6 \\ 10 & 14 \\ 18 & 22 \end{bmatrix}$$

MULTIPLICATION OF MATRICES

Multiplication of matrices is done *row by column*. If the dimensions of matrix **A** is M rows and N columns, then the dimensions of matrix **B** must have N rows. For example, given

DIMENSION $A(M, N), B(N, P), C(M, P)$.

if matrix **C** = matrix **A** $\times$ matrix **B**, then the dimension of **C** will be M rows and P columns.

EXAMPLE 1

To multiply $A(3, 2)$ and $B(2, 3)$:

Solution:

$$A = \begin{bmatrix} A(1,1) & A(1,2) \\ A(2,1) & A(2,2) \\ A(3,1) & A(3,2) \end{bmatrix} \qquad B = \begin{bmatrix} B(1,1) & B(1,2) & B(1,3) \\ B(2,1) & B(2,2) & B(2,3) \end{bmatrix}$$

$C(1, 1) = $ row 1 $\times$ column 1

$\qquad = A(1, 1) * B(1, 1) + A(1, 2) * B(2, 1)$

$C(1, 2) = $ row 1 $\times$ column 2

$\qquad = A(1, 1) * B(1, 2) + A(1, 2) * B(2, 2)$

$C(1, 3) = $ row 1 $\times$ column 3

$\qquad = A(1, 1) * B(1, 3) + A(1, 2) * B(2, 3)$

$$C(2, 1) = \text{row } 2 \times \text{column } 1$$
$$= A(2, 1) * B(1, 1) + A(2, 2) * B(2, 1)$$
$$C(2, 2) = \text{row } 2 \times \text{column } 2$$
$$= A(2, 1) * B(1, 2) + A(2, 2) * B(2, 2)$$
$$C(2, 3) = \text{row } 2 \times \text{column } 3$$
$$= A(2, 1) * B(1, 3) + A(2, 2) * B(2, 3)$$
$$C(3, 1) = \text{row } 3 \times \text{column } 1$$
$$= A(3, 1) * B(1, 1) + A(3, 2) * B(2, 1)$$
$$C(3, 2) = \text{row } 3 \times \text{column } 2$$
$$= A(3, 1) * B(1, 2) + A(3, 2) * B(2, 2)$$
$$C(3, 3) = \text{row } 3 \times \text{column } 3$$
$$= A(3, 1) * B(1, 3) + A(3, 2) * B(2, 3)$$

DIMENSION   C(3, 3)

EXAMPLE 2

Given the matrices

$$A = \begin{bmatrix} 2 & 7 \\ 5 & 4 \\ 3 & 6 \end{bmatrix} \qquad B = \begin{bmatrix} 1 & 3 & 5 \\ 2 & 4 & 6 \end{bmatrix}$$

Solution   Find matrix C = A $\times$ B.

DIMENSION   A(3, 2), B(2, 3)

Match in column and row as follows:

$$C(1, 1) = 2 * 1 + 7 * 2 = 2 + 14 = 16$$
$$C(1, 2) = 2 * 3 + 7 * 4 = 6 + 28 = 34$$
$$C(1, 3) = 2 * 5 + 7 * 6 = 10 + 42 = 52$$

$$C(2, 1) = 5 * 1 + 4 * 2 = 5 + 8 = 13$$
$$C(2, 2) = 5 * 3 + 4 * 4 = 15 + 16 = 31$$
$$C(2, 3) = 5 * 5 + 4 * 6 = 25 + 24 = 49$$

$$C(3, 1) = 3 * 1 + 6 * 2 = 3 + 12 = 15$$
$$C(3, 2) = 3 * 3 + 6 * 4 = 9 + 24 = 33$$
$$C(3, 3) = 3 * 5 + 6 * 6 = 15 + 36 = 51$$

$$C = \begin{bmatrix} 16 & 34 & 52 \\ 13 & 31 & 49 \\ 15 & 33 & 51 \end{bmatrix}$$

DIMENSION   C(3, 3)

EXERCISES

1. Find the dimensions of each matrix.

(a) $\begin{bmatrix} 5 & 4 & 3 & 2 \\ 7 & 6 & 8 & 3 \\ 4 & 1 & 4 & 2 \end{bmatrix}$

(b) $\begin{bmatrix} 3 & 4 \\ 4 & 5 \end{bmatrix}$

(c) $\begin{bmatrix} 8 & 11 \\ 7 & 12 \\ 12 & 41 \\ 14 & 32 \end{bmatrix}$

(d) $\begin{bmatrix} 24 & 37 \\ 35 & 86 \\ 46 & 92 \end{bmatrix}$

(e) $\begin{bmatrix} 1 & 0 & 8 & 4 & 3 & 2 \\ 6 & 3 & 4 & 5 & 7 & 9 \end{bmatrix}$

(f) $\begin{bmatrix} 27 & 98 & 69 \\ 42 & 87 & 142 \\ 86 & 77 & 29 \\ 99 & 29 & 100 \\ 45 & 72 & 141 \end{bmatrix}$

2. Given

$$X = \begin{bmatrix} 98 & 74 & 83 \\ 75 & 69 & 89 \\ 96 & 93 & 79 \\ 100 & 95 & 65 \\ 88 & 80 & 60 \end{bmatrix}$$

identify:

(a) Dimension of X (b) X(1, 3)
(c) X(2, 2) (d) X(3, 1)
(e) X(4, 2) (f) X(5, 3)
(g) X(5, 1)

3. Identify the elements of matrix **B** if **B = A**.

$$A = \begin{bmatrix} 25 & 43 & 71 & 81 \\ 72 & 59 & 53 & 93 \\ 68 & 80 & 62 & 69 \end{bmatrix}$$

(a) B(1, 1)   (b) B(1, 2)
(c) B(1, 4)   (d) B(2, 1)
(e) B(2, 2)   (f) B(2, 3)
(g) B(2, 4)   (h) B(3, 1)
(i) B(3, 3)   (j) B(3, 4)

4. Given

$$X = \begin{bmatrix} 17 & 26 & 35 \\ 25 & 38 & 47 \\ 34 & 49 & 58 \end{bmatrix} \qquad Y = \begin{bmatrix} 10 & 12 & 16 \\ 9 & 15 & 27 \\ 4 & 21 & 34 \end{bmatrix}$$

find:
(a) X + Y   (b) X + X
(c) Y - X   (d) 5 * X
(e) 3 * Y   (f) 4 * X + 2 * Y
(g) (-2) * X + Y

5. Draw a flowchart for the algorithm to add two matrices.

6. Draw a flowchart with two loops, one for rows and one for columns, to test for equality of two matrices.

7. For each matrix dimensioned below, find the dimensions of matrix **C** if **C = A * B**.
(a) A(3, 4), B(4, 2)   (b) A(2, 3), B(3, 5)
(c) A(7, 8), B(8, 6)   (d) A(9, 4), B(4, 8)
(e) A(2, 5), B(5, 2)   (f) A(6, 8), B(8, 10)
(g) A(5, 4), B(4, 11)   (h) A(6, 3), B(3, 2)
(i) A(7, 8), B(8, 2)   (j) A(5, 6), B(6, 5)

8. Given

$$W = \begin{bmatrix} 3 & 2 & 8 \\ 4 & -1 & 7 \\ 5 & -2 & 6 \end{bmatrix} \qquad X = \begin{bmatrix} 5 & 4 & 7 \\ 0 & 3 & -3 \\ 2 & 1 & 9 \end{bmatrix}$$

find:
(a) W * X   (b) X * W
(c) Is the multiplication of matrices cummutative?

SPECIAL MATRICES

Square Matrix

A matrix where the number of rows equals the number of columns is a *square matrix*.

$$A = \begin{bmatrix} 2 & 6 & 7 \\ 4 & 8 & 5 \\ 5 & 9 & 2 \end{bmatrix}$$

A is a square matrix since it has 3 rows and 3 columns.

Identity Matrix

The *identity matrix* is a square matrix with a main diagonal of 1's, and 0's in all other locations.

$$I = \begin{bmatrix} 1 & 0 & 0 \\ 0 & 1 & 0 \\ 0 & 0 & 1 \end{bmatrix}$$

The main diagonal will be the element of the matrix with equal subscripts.

Row subscript = column subscript

$$I(1, 1) = 1$$
$$I(2, 2) = 1$$
$$I(3, 3) = 1$$

The matrix I is the identity matrix because $A * I = A$.

$$A = \begin{bmatrix} 2 & 6 \\ 4 & 8 \end{bmatrix} \qquad \begin{bmatrix} 1 & 0 \\ 0 & 1 \end{bmatrix}$$

If $C = A * I$,

$$C(1, 1) = 2 * 1 + 6 * 0 = 2$$
$$C(1, 2) = 2 * 0 + 6 * 1 = 6$$
$$C(2, 1) = 4 * 1 + 8 * 0 = 4$$
$$C(2, 2) = 4 * 0 + 8 * 1 = 8$$

$$C = \begin{bmatrix} 2 & 6 \\ 4 & 8 \end{bmatrix} \qquad \text{which is matrix } \mathbf{A}$$

Inverse Matrices

Under certain conditions, an inverse matrix can exist. If, for example, the determinant of a matrix is zero, no inverse matrix can exist. When an inverse matrix does exist, it is written as matrix $\mathbf{A}^{-1}$ and $\mathbf{A} * \mathbf{A}^{-1} = \mathbf{I}$. That is, matrix $\mathbf{A}$, times its inverse, $\mathbf{A}^{-1}$, equals the identity matrix $\mathbf{I}$.

The determinant of the matrix with DIMENSION A(2, 2) is

$$\mathrm{DET}(A) = A(1, 1) * A(2, 2) - A(1, 2) * A(2, 1)$$

If $\mathrm{DET}(A) \neq 0$, then matrix $\mathbf{A}$ has an inverse. For example,

$$\mathbf{A} = \begin{bmatrix} 5 & 3 \\ 7 & 8 \end{bmatrix}$$

$$\mathrm{DET}(A) = 5 * 8 - 3 * 7$$
$$= 40 - 21$$
$$= 19$$

$\mathbf{A}$ has an inverse.

The inverse of $\mathbf{A}$,

$$\mathbf{A}^{-1} = \begin{bmatrix} a & b \\ c & d \end{bmatrix}$$

and $\mathbf{A} * \mathbf{A}^{-1} = \mathbf{I}$

$$a = A(2, 2)/\mathrm{DET}(A)$$
$$b = -A(1, 2)/\mathrm{DET}(A)$$
$$c = -A(2, 1)/\mathrm{DET}(A)$$
$$d = A(1, 1)/\mathrm{DET}(A)$$

For

$$\mathbf{A} = \begin{bmatrix} 5 & 3 \\ 7 & 8 \end{bmatrix}$$

$$\mathrm{DET}(A) = 19$$

$$A^{-1} = \begin{bmatrix} \frac{8}{19} & -\frac{3}{19} \\ -\frac{7}{19} & \frac{5}{19} \end{bmatrix}$$

$$A \times A^{-1} = \begin{bmatrix} 5 & 3 \\ 7 & 8 \end{bmatrix} * \begin{bmatrix} \frac{8}{19} & -\frac{3}{19} \\ -\frac{7}{19} & \frac{5}{19} \end{bmatrix}$$

$I(1, 1) = 5 * 8/19 + 3 * -7/19 = 40/19 - 21/19 = 19/19 = 1$

$I(1, 2) = 5 * -3/19 + 3 * 5/19 = -15/19 = 0$

$I(2, 1) = 7 * 8/19 + 8 * -7/19 = 56/19 + -56/19 = 0$

$I(2, 2) = 7 * -3/19 + 8 * 5/19 = -21/19 + 40/19 = 1$

$$A * A^{-1} = \begin{bmatrix} 1 & 0 \\ 0 & 1 \end{bmatrix} = I$$

For example, to find the inverse of

$$B = \begin{bmatrix} 12 & 10 \\ 18 & 15 \end{bmatrix}$$

$$DET(B) = 12 * 15 - 10 * 18$$

$$= 180 - 180$$

$$= 0$$

Matrix **B** does not have an inverse.

Transpose of a Matrix

The transpose of a matrix is another matrix with *rows and columns interchanged.* If $A(3, 2)$, its transpose is $A^T (2, 3)$. For example, given

$$C = \begin{bmatrix} 1 & 2 & 8 & 9 \\ 5 & 3 & 4 & 5 \\ 4 & 7 & 2 & 6 \end{bmatrix}$$

the transpose of **C** is C^T:

$$C^T = \begin{bmatrix} 1 & 5 & 4 \\ 2 & 3 & 7 \\ 8 & 4 & 2 \\ 9 & 5 & 6 \end{bmatrix}$$

DIMENSION $C(3, 4), C^T (4, 3)$

EXERCISES

1. Given

$$A = \begin{bmatrix} 7 & 13 & 26 \\ 9 & 14 & 32 \\ 11 & 17 & 43 \end{bmatrix}$$

   (a) Show that $A * I = A$.
   (b) Show that $I * A = A$.

2. Find the determinant of each matrix.

   (a) $\begin{bmatrix} 4 & 3 \\ 7 & 2 \end{bmatrix}$ (b) $\begin{bmatrix} 25 & 23 \\ 17 & 20 \end{bmatrix}$ (c) $\begin{bmatrix} 24 & 6 \\ 36 & -9 \end{bmatrix}$

   (d) $\begin{bmatrix} 128 & 48 \\ 75 & 42 \end{bmatrix}$ (e) $\begin{bmatrix} 125 & 80 \\ 50 & 32 \end{bmatrix}$

3. Find the inverse of each matrix in Exercise 2 if it exists.

4. Given

$$E = \begin{bmatrix} 15 & 12 \\ 21 & 8 \end{bmatrix}$$

   (a) Find DET(E).   (b) Find E^{-1}.
   (c) Show that $E * E^{-1} = I$ (d) Show that $E^{-1} * E = I$.

5. Find the dimensions of the transpose of each matrix.

   (a) $\begin{bmatrix} 14 & 88 \\ 23 & 93 \\ 42 & 75 \\ 56 & 68 \end{bmatrix}$ (b) $\begin{bmatrix} 1 & 2 & 4 & 6 & 8 \\ 3 & 6 & 12 & 18 & 24 \end{bmatrix}$

   (c) $\begin{bmatrix} 1 & 2 & 3 \\ 2 & 4 & 6 \\ 3 & 6 & 9 \\ 4 & 8 & 12 \\ 5 & 10 & 15 \\ 6 & 12 & 18 \end{bmatrix}$

(d) $\begin{bmatrix} 5 & 10 & 15 \\ 20 & 25 & 30 \\ 35 & 40 & 45 \end{bmatrix}$ (e) $\begin{bmatrix} 1 & 0 & 1 & 0 \\ 2 & 1 & 2 & 1 \\ 3 & 2 & 3 & 2 \\ 4 & 3 & 4 & 3 \end{bmatrix}$

6. Find the transpose of each matrix in Exercise 5.

MATRIX AS A TWO-DIMENSIONAL ARRAY

The programmer refers to the matrix as a two-dimensional array. To use an array, a programmer will:

1. Input data, identifying location by subscripts
2. Manipulate data in the array using some matrix functions
3. Output data from an array

INPUT DATA

There are two methods for data input: self-generated data and data read into the array. In both cases, the subscripts are used to identify locations.

EXAMPLE 1

Generate array C such that:

$$\begin{bmatrix} 0 & 2 & 2 & 2 & 2 \\ 1 & 0 & 2 & 2 & 2 \\ 1 & 1 & 0 & 2 & 2 \\ 1 & 1 & 1 & 0 & 2 \\ 1 & 1 & 1 & 1 & 0 \end{bmatrix}$$

For the array:

If subscripts are equal, the element is 0.

If the column subscript is greater than the row subscript, the element is 2.

If the column subscript is less than the row subscript, the element is 1.

The flowchart is

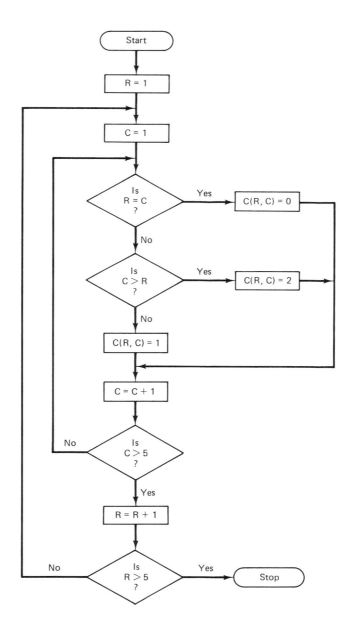

To trace the flowchart:

R	C	Is R = C?	Is C > R?	Is C > 5?	Is R > 5?	
1	1	Yes				C(1, 1) = 0
	2	No	Yes	No		C(1, 2) = 2
	3	No	Yes	No		C(1, 3) = 2
	4	No	Yes	No		C(1, 4) = 2
	5	No	Yes	No		C(1, 5) = 2
	6			Yes		
2					No	
	1	No	No	No		C(2, 1) = 1
	2	Yes				C(2, 2) = 0
	3	No	Yes	No		C(2, 3) = 2
	4	No	Yes	No		C(2, 4) = 2
	5	No	Yes	No		C(2, 5) = 2
	6			Yes		

(This continues until R = 6.)

Data can also be read into an array.

EXAMPLE 2

$$\text{Input array D} = \begin{bmatrix} 2.75 & 1.95 \\ 3.40 & 4.50 \\ 4.80 & 5.25 \end{bmatrix}$$

R	C	Is C > 2?	Is R > 3?	
1	1			D(1, 1) = 2.75
	2	No		D(1, 2) = 1.95
	3	Yes		
2			No	
	1			D(2, 1) = 3.40
	2	No		D(2, 2) = 4.50
	3	Yes		
3			No	
	1			D(3, 1) = 4.80
	2	No		D(3, 2) = 5.25
	3	Yes		
4			Yes	

The flowchart is

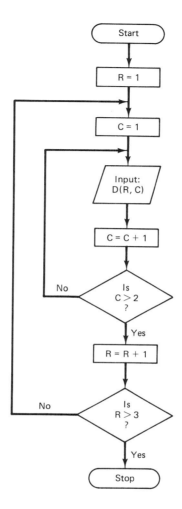

DATA MANIPULATION

Data manipulation usually involves addition of rows or columns.

For example: a survey was taken to determine public attitudes on a proposed flat rate tax bill. The results are listed in the following table:

Occupation	For	Against	No opinion
Education	257	198	22
Business	182	378	55
Clergy	82	148	12
Government	275	278	45
Retirees	158	197	58

A programmer now uses the raw survey data to tabulate the following:

1. Total responses in each occupational category
2. Total responses in each opinion category
3. Combined total responses for all occupational categories
4. Percentage by occupational category for each opinion
5. Total percentage by opinion category

1. To determine the total responses in each occupational category, calculate the sum of the row. There are five rows, so there will be five sums. The sum can become a fourth column in the array, which will now look like this:

Occupation	For	Against	No opinion	Sum
Education	————	————	————	————
Business	————	————	————	————
Clergy	————	————	————	————
Government	————	————	————	————
Retirees	————	————	————	————

The array, called SURVEY(R, 4), will contain the sum of each row.
The solution algorithm is:

Step 1:   R = 1 (Set row and column subscripts to 1)

Step 2:   C = 1

Step 3:   SURVEY (R, 4) = SURVEY (R, 4) + SURVEY (R, C) (Add the elements of a row and store in column 4)

Step 4:   C = C + 1 (Increment column subscript)

Step 5:   Is C > 3? (Test for final column subscript)
   No: Go to step 3
   Yes: Go to step 6

Step 6:   R = R + 1 (Increment row subscript)

Step 7:   Is R > 5? (Test for final row)
   No: Go to step 2
   Yes: Stop

2. To determine the total response in each opinion category, add another row and expand the array to six rows as shown here:

Occupation	For	Against	No opinion	Sum
Education	————	————	————	————
Business	————	————	————	————
Clergy	————	————	————	————
Government	————	————	————	————
Retirees	————	————	————	————
Sum	————	————	————	————

The array, SURVEY(6, C) will contain the sum of each column.

The solution is similar to the sum of the rows. The difference is that step 2 becomes

$$SURVEY(6, C) = SURVEY(6, C) + SURVEY(R, C)$$

3. The combined total responses for all occupational categories is either the sum of row 6 or the sum of column 4.

The solution can be to add the elements in row 6:

$$SURVEY(6, 4) = SURVEY(6, 4) + SURVEY(R, 4)$$

The first three calculations must be done before the percentage calculation.

4. To determine the percentage by occupational category for each opinion, use the ratio

$$\frac{\text{number in opinion category}}{\text{sum in occupational category}}$$

The "number in opinion category" will be the individual elements in each row. The "sum in occupational category" will be the elements in column 4 of each row.

The percent is the ratio times 100, since the ratio is a decimal. A ratio of .23 becomes 23%.

The solution is:

Step 1: R = 1 (Set row and column subscripts to 1)

Step 2: C = 1 (Set row and column subscripts to 1)

Step 3: (Calculate percentage PERCENT = SURVEY (R, C)/ SURVEY (R, 4) $*$ 100

Step 4: Output results

Step 5: C = C + 1 (Increment column subscript)

Step 6:   Is C > 3? (Test for final column)
          No: Go to step 3
          Yes: Go to step 7
Step 7:   R = R + 1 (Increment row subscript)
Step 8:   Is R > 5? (Test for final row)
          No: Go to step 2
          Yes: Stop

5. To determine the total percentage by opinion category, use the ratio

$$\frac{\text{Sum of opinion columns}}{\text{Total number of responses}}$$

The solution is:

Step 1:   C = 1
Step 2:   Total Percent = SURVEY(6, C)/SURVEY(6, 4) * 100
Step 3:   Output Total Percent
Step 4:   C = C + 1
Step 5:   Is C > 3?
          No: Go to step 2
          Yes: Stop

After the program is completed, the final array will be

$$\begin{bmatrix} 257 & 198 & 28 & 483 \\ 182 & 378 & 55 & 615 \\ 82 & 148 & 12 & 242 \\ 275 & 278 & 45 & 598 \\ 158 & 197 & 58 & 413 \\ 954 & 1199 & 198 & 2351 \end{bmatrix}$$

SURVEY (2, 4) = 615   The total number of business people responding to the survey.

SURVEY (6, 3) = 198   Of the total surveyed, 198 had no opinion.

SURVEY (6, 4) = 2351   Total number of responses.

EXERCISES

1. Use the following flowchart to find:
   (a) The dimension of B.
   (b) Each element in B.

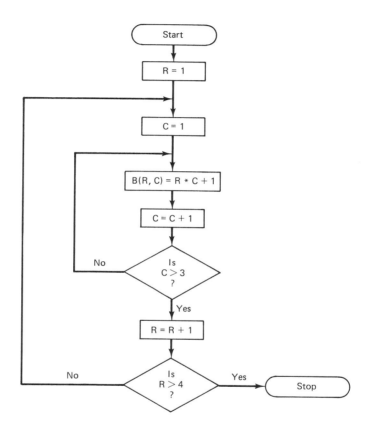

2. Draw a flowchart to generate this array.

$$\begin{bmatrix} 1 & 0 & 0 & 0 \\ 1 & 1 & 0 & 0 \\ 1 & 1 & 1 & 0 \\ 1 & 1 & 1 & 1 \end{bmatrix}$$

3. Draw a flowchart to generate this array.

$$\begin{bmatrix} 1 & 2 & 3 & 4 & 5 \\ 2 & 4 & 6 & 8 & 10 \\ 3 & 6 & 9 & 12 & 15 \\ 4 & 8 & 12 & 16 & 20 \\ 5 & 10 & 15 & 20 & 25 \\ 6 & 12 & 18 & 24 & 30 \end{bmatrix}$$

4. For the array VOTE:

Class	Yes	No
1	120	80
2	75	82
3	83	64
4	59	76

Draw a flowchart to:
(a) Add the sum of each row and store in VOTE(R, 3).
(b) Add the sum of each column and store the result in VOTE(5, C).
(c) Find the total responses for all four classes.

5. A survey was taken to determine the level of interest in various topics for a proposed meeting. Following are the tabulated results stored in a two-dimensional array called INT.

		Interest		
Topic	High	Medium	Low	No
1. Alternate modes of the successful dealer	25	23	10	1
2. Sales, new equipment	29	19	13	1
3. Sales, used equipment	27	24	8	1
4. Retail involvement of the manufacturer	24	16	18	2

Draw a flowchart to calculate the interest level of the participants by weighting the responses. Use the following formula and store the results in INT(5, 5).

H = number of high responses

M = number of medium responses

L = number of low responses

N = number of no responses

interest level = (3 * H + 2 * M + 1 * L + 0 * N)/(H + M + L + N)

SUMMARY

The mathematician sees a grid of numbers and recognizes a matrix. The programmer sees a grid of numbers and recognizes a two-dimensional array, which is useful in storing numbers in two categories and manipulating data.

The matrix has a dimension specified as row × column. These dimensions transfer to an array. The individual elements are identified by subscripts as A(I, J), where I respresents the row and J represents the column.

Matrices are equal if each element in one matches each element in the other. The operations for matrices are:

Addition:	Add individual elements.
Subtraction:	Subtract individual elements.
Multiply by a number K:	Each element is multiplied by K.

Multiplication of matrices is a row-by-column process. Therefore, to multiply matrices **A** and **B**—if the dimension of **A** is A(4, 3)—then **B** must have three rows. There can be any number of columns.

Special matrices include:

Square matrix:	Number of rows equals the number of columns.
Identity matrix:	In a square matrix if subscript for row equals the subscript for column, the element is 1. Otherwise, the element is 0.
Inverse matrix:	The inverse exists if the determinant $\neq 0$.
Transpose matrix:	Exchange rows with column elements. The dimension changes from A(I, J) to A(J, I).

The programmer utilizes the matrix as a two-dimensional array. The following then becomes important:

Dimension of array (row, column)
Identifying individual elements
Being able to find sums of rows and columns

The array is used for storage, output, and manipulation of data.

CHAPTER REVIEW

Given the matrix

$$D = \begin{bmatrix} 2 & 5 & 6 & 9 \\ 3 & 7 & 8 & 11 \\ 4 & 9 & 10 & 13 \end{bmatrix}$$

1. There are _____ rows.
2. There are _____ columns.
3. Elements $D(1, 2)$ = _____, and $D(2, 1)$ = _____.
4. The D^T has dimension $D^T ($ _____, _____ $)$.
5. If matrix

$$E = \begin{bmatrix} 1 & 2 \\ 3 & 4 \\ 5 & 6 \end{bmatrix}$$

is equal to matrix F, then

$$F(1, 1) = \text{\_\_\_\_}$$

$$F(1, 2) = \text{\_\_\_\_}$$

$$F(2, 1) = \text{\_\_\_\_}$$

$$F(2, 2) = \text{\_\_\_\_}$$

$$F(3, 1) = \text{\_\_\_\_}$$

$$F(3, 2) = \text{\_\_\_\_}$$

6. To add two matrices, the dimensions must be _____.

7. For matrices

$$B = \begin{bmatrix} 2 & 7 \\ 9 & 12 \end{bmatrix} \quad \text{and } A = \begin{bmatrix} 9 & 4 \\ 2 & 1 \end{bmatrix}$$

(a) $A + B = \begin{bmatrix} & \\ & \end{bmatrix}$

(b) $A - B = \begin{bmatrix} & \\ & \end{bmatrix}$

(c) $3 \times A = \begin{bmatrix} & \\ & \end{bmatrix}$

(d) $(-2) \times B = \begin{bmatrix} & \\ & \end{bmatrix}$

8. To multiply matrix **C** by matrix **D**—when DIM C(5, 4)—then for **D**, the rows must be _____ and the columns must be _____.

9. If matrices **A** and **B** are multiplied and DIM A(M, N), and DIM B(N, R), the product matrix **C** has DIM C(_____ , _____).

10. A square matrix with five rows has _____ columns.

11. If **I** has DIM I(4, 4) and is the identity matrix, then I(2, 2) = _____ and I(3, 2) = _____ .

12. Matrix **A** has an inverse if the _____ .

13. A matrix times its inverse equals _____ .

14. The transpose of matrix **A** with DIM A(5, 2) is matrix **A**^{T} with DIM **A**^{T}(_____ , _____).

15. A matrix to a programmer is a _____ array.

16. The two-dimensional array is used to _____ _____ data.

Number Base Concepts

chapter 17

The number system we commonly use is a base 10 system, or the decimal number system. But a number system can be based on other numbers besides 10. In fact, a number system can be based on any positive number greater than 1.

Any number system, whether it is a base 10 or base N, operates on the same principle. For some, familiarity with the base 10 system sometimes means not understanding the concept, but knowing the procedure. This chapter deals with base N systems where applying the procedures will stress the concepts.

BASE 10 NUMBERS

A number base uses the powers of the base. Our base 10 system uses powers of 10. The powers of 10 are

$$10^0 = 1 \qquad \text{(one)}$$
$$10^1 = 10 \qquad \text{(ten)}$$
$$10^2 = 100 \qquad \text{(hundred)}$$
$$10^3 = 1000 \qquad \text{(thousand)}$$
$$10^4 = 10000 \qquad \text{(ten thousand)}$$
$$10^5 = 100000 \qquad \text{(hundred thousand)}$$
$$10^6 = 1000000 \qquad \text{(million)}$$

The negative powers of 10 are

$$10^{-1} = .1 \qquad \text{(tenth)}$$
$$10^{-2} = .01 \qquad \text{(hundredth)}$$
$$10^{-3} = .001 \qquad \text{(thousandth)}$$
$$10^{-4} = .0001 \qquad \text{(ten thousandth)}$$
$$10^{-5} = .00001 \qquad \text{(hundred thousandth)}$$
$$10^{-6} = .000001 \qquad \text{(one millionth)}$$

A number such as 4187.2 is read as

"4 thousand 1 hundred 8 tens 7 ones and 2 tenths"

This number written in power form is

$$4 \times 10^3 + 1 \times 10^2 + 8 \times 10^1 + 7 \times 10^0 + 2 \times 10^{-1}$$

The number is really the sum of

$$4000 + 100 + 80 + 7 + .2$$

The place value of any digit is determined by its position in the number. To illustrate, the place values of base 10 are

$10^4 = 10000$	$10^3 = 1000$	$10^2 = 100$	$10^1 = 10$	$10^0 = 1$	.	$10^{-1} = 0.1$	$10^{-2} = 0.01$	$10^{-3} = 0.001$	$10^{-4} = 0.0001$

BASE 5 NUMBERS

The place value concept applies to any number base. A number in base 5 such as

(23) base 5

represents a place value of multiples of 5.

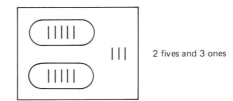

2 fives and 3 ones

Counting in base 5 would mean a group of five as a carry.

Base 5 Count

                0
                1
                2
                3
                4
               ——
               10   (first group of five)
               11
               12
               13
               14
               ——
               20   (second group of five)
               21
               22
               23
               24

Because a group of five means a carry into the next column, the only possible digits for a base 5 number will be the set

$$\{0, 1, 2, 3, 4\}$$

A base 5 number will have its place value determined by powers of 5. The powers of 5 are

$$5^0 = 1 \qquad\qquad 5^{-1} = .2$$
$$5^1 = 5 \qquad\qquad 5^{-2} = .04$$
$$5^2 = 25 \qquad\qquad 5^{-3} = .008$$
$$5^3 = 125 \qquad\qquad 5^{-4} = .0016$$
$$5^4 = 625 \qquad\qquad 5^{-5} = .00032$$
$$5^5 = 3125 \qquad\qquad 5^{-6} = .000064$$
$$5^6 = 15625$$

A number in base 5 such as

$$(231.42) \text{ base } 5$$

has a place value of

$$2 \times 5^2 + 3 \times 5^1 + 1 \times 5^0 + 4 \times 5^{-1} + 2 \times 5^{-2}$$

PLACE VALUE IN BASE *N*

Any number base can be expanded in the same way. For example, base 3 uses a 3 for a carry, so a number in base 3 will have as its set of digits

$$\{0, 1, 2\}$$

EXAMPLE 1

Find the place value of the following base 3 number:

$$(2122.212) \text{ base } 3$$

Solution　Use powers of 3 and write the number in expanded form.

$$2 \times 3^3 + 1 \times 3^2 + 2 \times 3^1 + 2 \times 3^0 + 2 \times 3^{-1} + 1 \times 3^{-2} + 2 \times 3^{-3}$$

The place value of each digit depends on the power of 3.

$$3^0 = 1 \qquad\qquad 3^{-1} = .33\overline{3}$$
$$3^1 = 3 \qquad\qquad 3^{-2} = .11\overline{1}$$
$$3^2 = 9 \qquad\qquad 3^{-3} = .\overline{037}$$
$$3^3 = 27$$

EXAMPLE 2

Find the place value of the base 6 number.

$$(4513.14) \text{ base } 6$$

Solution   The set of digits in base 6 is

$$\{0, 1, 2, 3, 4, 5\}$$

$$4 \times 6^3 + 5 \times 6^2 + 1 \times 6^1 + 3 \times 6^0 + 1 \times 6^{-1} + 4 \times 6^{-2}$$

The place value of each digit is

$$6^3 \ = 216$$
$$6^2 \ = 36$$
$$6^1 \ = 6$$
$$6^0 \ = 1$$
$$6^{-1} = .166\overline{6}$$
$$6^{-2} = .027\overline{7}$$

BASE *N* TO BASE 10

To change a base N number to base 10:

1. Write the number in power form.
2. Multiply digits by the place value.
3. Add the products.

EXAMPLE 1

Change (132.312) base 4 to base 10.

Solution:

1. Number in power form:

$$1 \times 4^2 + 3 \times 4^1 + 2 \times 4^0 + 3 \times 4^{-1} + 1 \times 4^{-2} + 2 \times 4^{-3}$$

Powers of 4 are:

$$4^2 \ = 16$$
$$4^1 \ = 4$$
$$4^0 \ = 1$$
$$4^{-1} = .25$$
$$4^{-2} = .0625$$
$$4^{-3} = .015625$$

2.   Multiply digits by place value:

$$1 \times 16 = 16.$$
$$3 \times 4 = 12.$$
$$2 \times 1 = \ \ 2.$$
$$3 \times .25 = \quad .75$$
$$1 \times .0625 = \quad .0625$$
$$2 \times .015625 = \quad .03125$$

3.   Sum of the products: 30.84375

(132.312) base 4 = 30.84375

EXAMPLE 2

Using base 7, change (1632.12) base 7 to base 10.

Solution:

1.   Number in power form:

$$1 \times 7^3 + 6 \times 7^2 + 3 \times 7^1 + 2 \times 7^0 + 1 \times 7^{-1} + 2 \times 7^{-2}$$

Powers of 7 are:

$$7^3 \ = 343$$
$$7^2 \ = 49$$
$$7^1 \ = 7$$
$$7^0 \ = 1$$
$$7^{-1} = .1428571$$
$$7^{-2} = .0204082$$

2.   Multiply digits by place value:

$$1 \times 343 = 343.$$
$$6 \times 49 = 294.$$
$$3 \times 7 = \ \ 21.$$
$$2 \times 1 = \ \ \ 2.$$
$$1 \times .1428571 = \quad .1428571$$
$$2 \times .0204082 = \quad .0408164$$

3.   Sum of the products: 658.1836735

(1632.12) base 7 = 658.1836735

BASE 10 TO BASE *N*

There are two methods to determine the base *N* value of a base 10 number.

Method I

Divide the place values of the base into the number and its remainders until the remainder is zero.

EXAMPLE 1

Change 52 to base 3.

Solution   Find the place values of 3 that are less than 52.

$$3^0 = 1$$
$$3^1 = 3$$
$$3^2 = 9$$
$$3^3 = 27$$
$$3^4 = 81 \qquad \text{(this is greater than 52, so the first divisor is 27)}$$

$$
\begin{array}{llll}
\;\;\;\;1 & \;\;\;\;2 & \;\;\;\;2 & \;\;\;\;1 \\
27\,\overline{)52} & 9\,\overline{)25} & 3\,\overline{)7} & 1\,\overline{)1} \\
\;\;\;\underline{27} & \;\;\;\underline{18} & \;\;\;\underline{6} & \;\;\;\underline{1} \\
\;\;\;\textcircled{25} & \;\;\;\textcircled{7} & \;\;\;\textcircled{1} & \;\;\;\;0
\end{array}
$$

(remainder is zero; the calculation is complete)

$$52 = (1221) \text{ base } 3$$

To check, change (1221) base 3 to base 10:

$$1 \times 3^3 + 2 \times 3^2 + 2 \times 3^1 + 1 \times 3^0$$
$$1 \times 27 = 27$$
$$2 \times 9 = 18$$
$$2 \times 3 = \;\;6$$
$$1 \times 1 = \;\;\underline{1}$$
$$52$$

EXAMPLE 2

Change 198 to base 5.

Solution   Find the powers of 5:

$$5^0 = \;\;1$$
$$5^1 = \;\;5$$
$$5^2 = \;\;25$$
$$5^3 = 125$$
$$5^4 = 625 \qquad \text{(stop at this power because } 625 > 198 \text{)}$$

Divide:

$$
\begin{array}{cccc}
1 & 2 & 4 & 3 \\
125\,\overline{)198} & 25\,\overline{)73} & 5\,\overline{)23} & 1)\overline{3} \\
\underline{125} & \underline{50} & \underline{20} & \underline{3} \\
73 & 23 & 3 & 0
\end{array}
$$

198 = (1243) base 5.

Method II

Divide by the base until the quotient is 0. The number is the remainders.

EXAMPLE 3

Using method II, change 386 to base 7.

Solution:

$$
\begin{array}{ll}
7\ \underline{|\,386} & \\
7\ \ \underline{|\,55} & \text{remainder } 1 \ \ \uparrow \\
7\ \ \ \ \underline{|\,7} & \text{remainder } 6 \\
7\ \ \ \ \ \underline{|\,1} & \text{remainder } 0 \\
\ \ \ \ \ \ 0 & \text{remainder } 1
\end{array}
$$

386 = (1061) base 7

Check: (1061) base 7

$$1 \times 7^3 + 0 \times 7^2 + 6 \times 7^1 + 1 \times 7^0$$

$$1 \times 343 = 343$$

$$0 \times 49 = 0$$

$$6 \times 7 = 42$$

$$1 \times 1 = \underline{1}$$

$$\text{Sum:} \quad 386$$

EXAMPLE 4

Using method II, change 4594 to base 9.

Solution:

$$
\begin{array}{ll}
9\ \underline{|\,4594} & \\
9\ \ \underline{|\,510} & \text{remainder } 4 \ \ \uparrow \\
9\ \ \ \underline{|\,56} & \text{remainder } 6 \\
9\ \ \ \ \ \underline{|\,6} & \text{remainder } 2 \\
\ \ \ \ \ \ 0 & \text{remainder } 6
\end{array}
$$

4594 = (6264) base 9

EXAMPLE 5

Using method I, change 4594 to base 9.

Solution:

$$9^0 = 1$$
$$9^1 = 9$$
$$9^2 = 81$$
$$9^3 = 729$$
$$9^4 = 6561 \qquad (6561 > 4594)$$

$$
\begin{array}{cccc}
6 & 2 & 6 & 4 \\
729\overline{)4594} & 81\overline{)220} & 9\overline{)58} & 1\overline{)4} \\
\underline{4374} & \underline{162} & \underline{54} & \underline{4} \\
22 & 58 & 4 & 0
\end{array}
$$

4594 = (6264) base 9

EXERCISES

1. Write in power form.
   (a) 578.9322   (b) (1321.221) base 4
   (c) (43.422) base 5   (d) (178.77) base 9
   (e) (2212.222) base 3
2. Write the first 10 numbers of base 3 starting with 1.
3. Identify the base and the number represented by each diagram.

(a)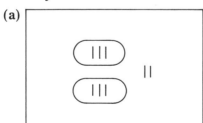

(b)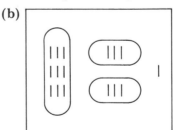

(c)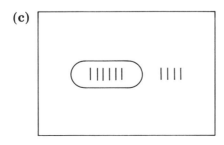

(d)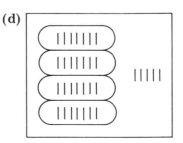

(e)

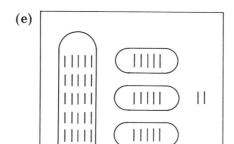

4. Identify the set of digits for each number base.
   (a) Base 6   **(b)** Base 4
   (c) Base 2   **(d)** Base 8
   (e) Base 9

5. Complete the table.

	$N = 10$	$N = 8$	$N = 7$	$N = 6$	$N = 3$
N^5					
N^4					
N^3					
N^2					
N^1					
N^0					
N^{-1}					
N^{-2}					
N^{-3}					
N^{-4}					

6. Change into base 10.
   (a) (1212) base 3   **(b)** (157) base 9
   (c) (101111) base 2   **(d)** (3231) base 4
   (e) (4132) base 5   **(f)** (51234) base 6
   (g) (3654) base 7   **(h)** (3746) base 8
   (i) (211021) base 3   **(j)** (23002131) base 4

7. Using Method I, change 378 to the base specified.
   (a) Base 3   **(b)** Base 6
   (c) Base 5   **(d)** Base 9
   (e) Base 2

8. Using Method II, change 1529 to the base specified.
   (a) Base 2   **(b)** Base 8
   (c) Base 4   **(d)** Base 3
   (e) Base 7

CHANGING DECIMALS TO BASE *N*

The method for converting decimal numbers is different from the methods used to convert whole numbers.

EXAMPLE 1

Change 0.5432 to base 5.

Solution   Multiply by 5.

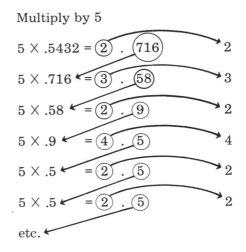

(At this point the decimal is repeating in base 5.)

$$0.5432 = (.232422\overline{2}) \text{ base } 5$$

EXAMPLE 2

Change .145 to base 9.

Solution:

$$9 \times .145 = 1.305 \qquad 1$$
$$9 \times .305 = 2.745 \qquad 2$$
$$9 \times .745 = 6.705 \qquad 6$$
$$9 \times .705 = 6.345 \qquad 6$$
$$9 \times .345 = 3.105 \qquad 3$$
$$9 \times .105 = 0.945 \qquad 0$$
$$9 \times .945 = 8.505 \qquad 8$$
$$9 \times .505 = 4.545 \qquad 4$$

$$9 \times .545 = 4.905 \qquad 4$$
$$9 \times .905 = 8.145 \qquad 8$$
$$\text{etc.}$$

$$.145 = (.1266308448 \ldots) \text{ base } 9$$

A terminating decimal may be nonterminating in another base.

EXAMPLE 3

Change 75.8 to base 4.

Solution:

Change 75 to base 4	Change .8 to base 4
4 ⎸75	$4 \times .8 = 3.2 \qquad 3$
4 ⎸18 R 3	$4 \times .2 = 0.8 \qquad 0$
4 ⎸4 R 2	$4 \times .8 = 3.2 \qquad 3$
4 ⎸1 R 0	$4 \times .2 = 0.8 \qquad 0$
0 R 1	etc.

$$75.8 = (1023.3\overline{030}) \text{ base } 4$$

EXAMPLE 4

Change 943.75 to base 8.

Solution   First: Change 943:

$$8^0 = \quad 1$$
$$8^1 = \quad\ 8$$
$$8^2 = \quad 64$$
$$8^3 = \ 512$$
$$8^4 = 4096 \qquad (> 943)$$

$$\begin{array}{cccc}
1 & 6 & 5 & 7 \\
512\,\overline{)943} & 64\,\overline{)431} & 8\,\overline{)47} & 1\,\overline{)7} \\
\underline{512} & \underline{384} & \underline{40} & \underline{7} \\
431 & 47 & 7 & 0
\end{array}$$

$$943 = (1657) \text{ base } 8$$

Then change .75:

$$8 \times .75 = 6.000 \qquad 6$$
$$943.75 = (1657.6) \text{ base } 8$$

CONVERSION BETWEEN BASES

To convert (43.01) base 5 to base 9, use the following two steps:

1. Change from base 5 to base 10.
2. Change from base 10 to base 9.

Therefore, converting (43.01) base 5 to base 10:

$$4 \times 5^1 + 3 \times 5^0 + 0 \times 5^{-1} + 1 \times 5^{-2} =$$

$$20 \quad + \quad 3 \quad + \quad 0 \quad + \quad .04 \quad = 23.04$$

Next, convert 23.04 to base 9:

23 to base 9	.04 to base 9	
9 ⎣23	$9 \times .04 = 0.36$	0
9 ⎣2 R 5	$9 \times .36 = 3.24$	3
0 R 2	$9 \times .24 = 2.16$	2
	$9 \times .16 = 1.44$	1
	$9 \times .44 = 3.96$	3
	etc.	

$$23.04 = (25.03213 \ldots) \text{ base } 9$$

$$(43.01) \text{ base } 5 = (25.03213 \ldots) \text{ base } 9$$

In certain cases where the bases are powers of each other, conversion is easier.

EXAMPLE 1

Consider base 3 and base 9, where $9 = 3^2$. Change (12112.2111) base 3 to base 9.

Solution   To change from base 3 to base 9 is a direct change since 9 is 3^2. Starting at the decimal point, group the numbers by twos, and add zeros where necessary.

$$(\underline{0\ 1}\ \underline{2\ 1}\ \underline{1\ 2}\ .\ \underline{2\ 1}\ \underline{1\ 1}) \text{ base } 3$$

In grouping by twos, the numbers are changed to their base 9 value.

$$01 = 1$$

$$21 = 7 \qquad (2 \text{ threes} + 1 \text{ one})$$

$$12 = 5 \qquad (1 \text{ three} + 2 \text{ ones})$$

$$11 = 4 \qquad (1 \text{ three} + 1 \text{ one})$$

$$(\underbrace{0\,1} \ \underbrace{2\,1} \ \underbrace{1\,2} \ . \ \underbrace{2\,1} \ \underbrace{1\,1}) \text{ base 3}$$
$$(\quad 1 \qquad 7 \qquad 5 \ . \quad 7 \qquad 4 \quad) \text{ base 9}$$

Base 9 to base 3 reverses the process.

EXAMPLE 2

Change (478.763) base 9 to base 3.

Solution   Each digit is changed to a two-digit base 3 number.

Base 3

$$4 = 11 \qquad (1 \text{ three} + 1 \text{ one})$$

$$7 = 21 \qquad (2 \text{ threes} + 1 \text{ one})$$

$$8 = 22 \qquad (2 \text{ threes} + 2 \text{ ones})$$

$$6 = 20 \qquad (2 \text{ threes} + 0 \text{ ones})$$

$$3 = 10 \qquad (1 \text{ three} + 0 \text{ ones})$$

$$(\ 4 \quad 7 \quad 8 \ . \ 7 \quad 6 \quad 3 \ ) \text{ base 9}$$
$$(11 \ \ 21 \ \ 22 \ . \ 21 \ \ 20 \ \ 10) \text{ base 3}$$

$$(478.763) \text{ base 9} = (112122.212010) \text{ base 3}$$

This method works only for converting between bases that are powers of each other.

EXERCISES

1. Change .3875 to the bases specified.
    (a) Base 8   (b) Base 5
    (c) Base 3   (d) Base 7
    (e) Base 9

2. Change to the specified base.
   (a) 75.24 to base 5   (b) 428.1231 to base 9
   (c) 459.666̄ to base 3   (d) 16.4532 to base 6
   (e) 1.47893 to base 8

3. Change (13232.211) base 4 to the following base specified.
   (a) Base 5   (b) Base 8
   (c) Base 9   (d) Base 2
   (e) Base 7

4. Change (12210112.21121) base 3 to base 9.

5. Change (67845.374456) base 9 to base 3.

6. Change (11011.11001101) base 2 to base 4.

7. Change (32102001.212133) base 4 to base 2.

ADDITION IN BASE 5

When doing addition with base numbers, it is easier to use a table. To construct an addition table for a given base number, add the basic numbers in that base. For example, to construct an addition table for base 5, use the basic numbers, 0, 1, 2, 3, and 4, the only possible additives in base 5.

$$
\begin{array}{ccccc}
1 & 1 & 1 & 1 & 1 \\
+0 & +1 & +2 & +3 & +4 \\
\hline
1 & 2 & 3 & 4 & 10
\end{array}
\quad (1 + 4 = 5,\ 1\ \text{five} + 0\ \text{ones})
$$

$$
\begin{array}{ccccc}
2 & 2 & 2 & 2 & 2 \\
+0 & +1 & +2 & +3 & +4 \\
\hline
2 & 3 & 4 & 10 & 11
\end{array}
\quad (2 + 4 = 6,\ 1\ \text{five} + 1\ \text{one})
$$

$$
\begin{array}{ccccc}
3 & 3 & 3 & 3 & 3 \\
+0 & +1 & +2 & +3 & +4 \\
\hline
3 & 4 & 10 & 11 & 12
\end{array}
\quad (3 + 4 = 7,\ 1\ \text{five} + 2\ \text{ones})
$$

$$
\begin{array}{ccccc}
4 & 4 & 4 & 4 & 4 \\
+0 & +1 & +2 & +3 & +4 \\
\hline
4 & 10 & 11 & 12 & 13
\end{array}
\quad (4 + 4 = 8,\ 1\ \text{five} + 3\ \text{ones})
$$

The table in base 5 is:

+	0	1	2	3	4
0	0	1	2	3	4
1	1	2	3	4	10
2	2	3	4	10	11
3	3	4	10	11	12
4	4	10	11	12	13

Using the table, add (134) base 5 + (4132) base 5.

$$
\begin{array}{r}
1\ |\ 1\ | \\
(1\ |\ 3\ |\ 4)\ \text{base 5} \\
+(4\ 1\ |\ 3\ |\ 2)\ \text{base 5} \\
\hline
4\ 3\ |\ 2\ |\ 1 \leftarrow (4+2\ \text{on the addition table} = 11)
\end{array}
$$

(3 + 3 = 11 from the table + 1 carried = 12)

(1 + 1 = 2 plus 1 carried = 3)

(134) base 5 + (4132) base 5 = (4321) base 5

Without the table, addition can be done by remembering that a 5 is 10 in base 5. For example:

Add:
$$
\begin{array}{r}
1\ |\ 1\ | \\
(4\ |\ 3\ |\ 4)\ \text{base 5} \\
+\ (4\ |\ 1\ |\ 3)\ \text{base 5} \\
\hline
1\ 4\ |\ 0\ |\ 2 \leftarrow (3+4=7,\ 1\ \text{five} + 2\ \text{ones})
\end{array}
$$

(1 + 3 + 1 = 5, 1 five + 0 ones)

(4 + 4 + 1 = 9, 1 five + 4 ones)

ADDITION IN BASE *N*

To set up a table in any base, use the set of digits of that base. For example, for base 3, the set of digits is 0, 1, and 2. Set up the table using only the digits 0, 1, and 2 horizontally and vertically.

+	0	1	2	
0	0	1	2	$(0 + N = N)$
1	1	2	10	(for 1 + 2 = 10 in base 3)
2	2	10	11	(2 + 2 = 11, 1 three + 1 one)

Using the addition table above, add the following numbers:

$$
\begin{array}{r}
1\ 1 \\
(1\ 2\ 2\ 0\ 1)\ \text{base 3} \\
+(2\ 1\ 2\ 2\ 1)\ \text{base 3} \\
\hline
(11\ 1\ 1\ 2\ 2)\ \text{base 3}
\end{array}
$$

+	0	1	2
0	0	1	2
1	1	2	10
2	2	10	11

Add the following numbers without using a table.

$$
\begin{array}{r}
1\,|1\,|\ \ |1\,|1\,| \\
(2\,|1\,|3\,|0\,|2\,|3) \text{ base 4} \\
+(\ \ \ |3\,|2\,|2\,|1\,|2) \text{ base 4} \\
\hline
(3\,|1\,|1\,|3\,|0\,|1) \leftarrow (3 + 2 = 11, \text{ 1 four} + \text{1 one})
\end{array}
$$

$$(1 + 2 + 1 = 10, \text{ 1 four} + \text{0 ones})$$

$$(1 + 0 + 2 = 3, \text{ 0 fours} + \text{3 ones})$$

$$(3 + 2 = 11, \text{ 1 four} + \text{1 one})$$

$$(3 + 1 + 1 = 11, \text{ 1 four} + \text{1 one})$$

$$(2 + 1 = 3, \text{ 0 fours} + \text{3 ones})$$

SUBTRACTION IN BASE N

A problem in base 5, such as

$$3 + 4 = (12) \text{ base 5}$$

can become the two subtraction problems:

$$12 - 4 = 3 \qquad \text{and} \qquad 12 - 3 = 4$$

Here the addition table must be used to find the addend.
Subtract the following numbers in base 5:

$$
\begin{array}{r}
3\ \ |12\,|11\,|\ 1 \\
(4\,|3\,|2\,|2\,|4) \text{ base 5} \\
-(\ \ \ |3\,|2\,|4\,|2) \text{ base 5} \\
\hline
(3\,|4\,|4\,|3\,|2) \text{ base 5}
\end{array}
$$

+	0	1	2	3	4
0	0	1	2	3	4
1	1	2	3	4	10
2	2	3	(4)	10	11
3	3	4	10	11	12
4	4	10	11	12	13

$$4 - 2 = 2$$

+	0	1	2	3	4
0	0	1	2	3	4
1	1	2	3	4	10
2	2	3	4	10	11
3	3	4	10	11	(12)
4	4	10	11	12	13

$$12 - 4 = 3$$

+	0	1	2	3	4
0	0	1	2	3	4
1	1	2	3	4	10
2	2	3	4	10	11
3	3	4	10	11	12
4	4	10	(11)	12	13

11 – 2 = 4

+	0	1	2	3	4
0	0	1	2	3	4
1	1	2	3	4	10
2	2	3	4	10	11
3	3	4	10	11	12
4	4	10	11	(12)	13

12 – 3 = 4

To check:

$$(3\ 2\ 4\ 2)\text{ base 5}$$
$$+(3\ 4\ 4\ 3\ 2)\text{ base 5}$$
$$(4\ 3\ 2\ 2\ 4)\text{ base 5}$$

SUBTRACTION BY COMPLEMENTS

Subtraction can be accomplished through the addition of complements. The complement of a number N is formed by subtracting the digits of the number from the highest digit of that base $(N - 1)$. For example:

The complement of 4378 in base 10 is

$$\begin{array}{r} 9999 \\ -\,4378 \\ \hline 5621 \end{array}$$

(highest digits in base 10)

(9's complement of 4378)

The complement of 5463 in base 7 is

$$\begin{array}{r} 6666 \\ -\,5463 \\ \hline 1203 \end{array}$$

(highest digits in base 7)

(6's complement of 5463)

The complement of 21202 in base 4 is

$$\begin{array}{r} 33333 \\ -\,21202 \\ \hline 12131 \end{array}$$

(highest digits in base 4)

(3's complement of 21202)

The complement of 1011011 in base 2 is

$$\begin{array}{r} 1111111 \\ -\,1011011 \\ \hline 0100100 \end{array}$$

(highest digits in base 2)

(1's complement of 1011011)

Do the following subtraction:

$$(4\ 1\ 2\ 3\ 4\ 2)\ \text{base}\ 5$$
$$-\ (2\ 3\ 4\ 4\ 3\ 4)\ \text{base}\ 5$$
$$\overline{(1\ 2\ 2\ 4\ 0\ 3)\ \text{base}\ 5}$$

Using the 4's complement of (2 3 4 4 3 4) base 5, this subtraction can be done as an addition problem.

$$4\ 4\ 4\ 4\ 4\ 4$$
$$-\ (2\ 3\ 4\ 4\ 3\ 4)\ \text{base}\ 5$$
$$\overline{2\ 1\ 0\ 0\ 1\ 0} \qquad (\text{4's complement of } 234434)$$

This changes the subtraction to

$(4\ 1\ 2\ 3\ 4\ 2)\ \text{base}\ 5$	$(4\ 1\ 2\ 3\ 4\ 2)\ \text{base}\ 5$	
$-\ (2\ 3\ 4\ 4\ 3\ 4)\ \text{base}\ 5$	$+\ (2\ 1\ 0\ 0\ 1\ 0)$	(4's complement)
$(1\ 2\ 2\ 4\ 0\ 3)\ \text{base}\ 5$	$①\ 1\ 2\ 2\ 4\ 0\ 2$	
	$+\ 1$	(carry is added)
	$(1\ 2\ 2\ 4\ 0\ 3)\ \text{base}\ 5$	

Subtract using complements:

$(2\ 1\ 2\ 0\ 1\ 1)\ \text{base}\ 3$	$(2\ 1\ 2\ 0\ 1\ 1)\ \text{base}\ 3$	
$-\ (1\ 2\ 1\ 2\ 2\ 2)\ \text{base}\ 3$	$+\ 1\ 0\ 1\ 0\ 0\ 0$	(2's complement)
	$①\ 0\ 2\ 0\ 0\ 1\ 1$	
	$+\ 1$	(add the carry digit)
	$(2\ 0\ 0\ 1\ 2)\ \text{base}\ 3$	

To *check* the answer, use addition:

$$(1\ 2\ 1\ 2\ 2\ 2)\ \text{base}\ 3$$
$$+\ \ \ (2\ 0\ 0\ 1\ 2)\ \text{base}\ 3$$
$$\overline{(2\ 1\ 2\ 0\ 1\ 1)\ \text{base}\ 3}$$

As a rule, for subtraction using complements, follow these steps:

1. For base N, find the $(N-1)$ complement.
2. Add the complement.
3. Add the carry digit to the lowest place value.

EXAMPLE 1

Using complements, subtract:

$$(1\ 6\ 5\ 4\ 3)\ \text{base}\ 7$$
$$-\quad (5\ 6\ 5\ 4)\ \text{base}\ 7$$

Solution   In complement subtraction the digits of both numbers must match. To form the 6's complement, use the number (05654) so that the digits will match. To form the complement:

$$
\begin{array}{r}
6\ 6\ 6\ 6\ 6 \\
-\ 0\ 5\ 6\ 5\ 4 \\
\hline
6\ 1\ 0\ 1\ 2
\end{array}
\qquad \text{(6's complement of 05654)}
$$

To do the subtraction, add the complement.

$$
\begin{array}{r}
(1\ 6\ 5\ 4\ 3)\ \text{base}\ 7 \\
+\ 6\ 1\ 0\ 1\ 2 \qquad \text{(6's complement)} \\
\hline
①\,1\ 0\ 5\ 5\ 5 \\
+1 \\
\hline
1\ 0\ 5\ 5\ 6
\end{array}
$$

EXAMPLE 2

Using complements, subtract and check the results in base 4 for the following numbers:

$$(1\ 2\ 2\ 1\ 1)\ \text{base}\ 4$$
$$-\quad (2\ 3\ 3)\ \text{base}\ 4$$

Solution   Rewriting the number using the 0's to get matching digits, the problem becomes

$$(1\ 2\ 2\ 1\ 1)\ \text{base}\ 4 \qquad\qquad (1\ 2\ 2\ 1\ 1)\ \text{base}\ 4$$
$$(0\ 0\ 2\ 3\ 3)\ \text{base}\ 4 \qquad\qquad
\begin{array}{r}
+(3\ 3\ 1\ 0\ 0) \qquad \text{(3's complement)} \\
\hline
1\ 1\ 1\ 3\ 1\ 1 \\
+\ 1 \\
\hline
(1\ 1\ 3\ 1\ 2)\ \text{base}\ 4
\end{array}
$$

To check:

$$
\begin{array}{r}
(0\ 0\ 2\ 3\ 3)\ \text{base}\ 4 \\
+(1\ 1\ 3\ 1\ 2)\ \text{base}\ 4 \\
\hline
(1\ 2\ 2\ 1\ 1)\ \text{base}\ 4
\end{array}
$$

SUBTRACTION WITH NEGATIVE DIFFERENCES

Normally, subtraction with negative differences is done by subtracting a larger number from a smaller number. For example, in base 10,

$$
\begin{array}{r}
75 \\
-96 \\
\hline
-21
\end{array}
$$

However, subtraction by complements is different if the answer is a negative number.

$$
\begin{array}{r}
75 \\
-96 \\
\hline
-21
\end{array}
\qquad
\begin{array}{r}
75 \\
+03 \\
\hline
78
\end{array}
\quad
\begin{array}{l}
\text{(9's complement)} \\
\text{(no carry occurs)}
\end{array}
$$

If no carry occurs, the answer is a negative number and is the $(N-1)$ complement.

78 (9's complement)   which becomes -21 (base 10)

Remember, in subtraction with negative differences, if there is no carry, the answer is in complement form. Change the complement to base N and add a minus sign.

EXAMPLE 1

Subtract in base 3 using complements.

$$
\begin{array}{r}
(1\ 2\ 2\ 1)\ \text{base 3} \\
-\ (2\ 1\ 2\ 2)\ \text{base 3}
\end{array}
$$

$$
\begin{array}{r}
(1\ 2\ 2\ 1)\ \text{base 3} \\
+\ \ 0\ 1\ 0\ 0 \qquad \text{(2's complement)} \\
\hline
\text{(no carry)}\quad (2\ 0\ 2\ 1) \qquad \text{(2's complement)}
\end{array}
$$

Solution   Reverse the complement and add a minus sign: $-(0.201)$ base 3.

EXAMPLE 2

For the following problem, subtract and check the answer using complements:

$$
\begin{array}{r}
(3\ 4\ 2\ 1\ 1)\ \text{base 6} \\
-\ (4\ 3\ 1\ 3\ 3)\ \text{base 6}
\end{array}
$$

Solution:

$$(3\ 4\ 2\ 1\ 1)\ \text{base 6}$$
$$+(1\ 2\ 4\ 2\ 2)\qquad (5\text{'s complement})$$
no carry $\overline{(5\ 1\ 0\ 3\ 3)}\qquad (5\text{'s complement of negative answer})$

The answer is $-(04522)$ base 6.

    To check (the problem is still a subtraction since it is the addition of a negative and a positive number):

$$(4\ 3\ 1\ 3\ 3)\ \text{base 6}$$
$$+(5\ 1\ 0\ 3\ 3)\qquad (5\text{'s complement})$$
carry $\quad\textcircled{1}\ 3\ 4\ 2\ 1\ 0$
$$+\ 1$$
$\overline{(3\ 4\ 2\ 1\ 1)\ \text{base 6}}\qquad$ the answer checks

EXERCISES

1. Using a base 5 addition table, add:
   (a) (21) base 5 + (42) base 5
   (b) (420) base 5 + (344) base 5
   (c) (3143) base 5 + (24311) base 5
   (d) (12) base 5 + (24) base 5 + (43) base 5
   (e) (4321) base 5 + (3422) base 5 + (2434) base 5
2. Fill in the addition table for base 9.

+	0	1	2	3	4	5	6	7	8
0									
1									
2									
3									
4									
5									
6									
7									
8									

3. Using the table in Exercise 2, add:
   (a)   (6 4 5 8) base 9   (b)   (7 8 3 8 5) base 9
        +(5 5 6 8) base 9   +(8 5 6 4 7) base 9

4. Add.
   (a)   (2 1 1 0 2 2) base 3   (b)   (6 5 6 3 4) base 7
        +(1 2 2 2 1 2) base 3   +(4 2 5 6 3) base 7

   (c)   (1 5) base 6   (d)   (1 3 2 1 2 3 3) base 4
          (2 4) base 6   +(2 3 3 2 2 0 3) base 4
          (3 1) base 6
        +(5 5) base 6

5. Counting is the process of adding 1 to find each successive digit. Fill in the missing numbers for each count.
   (a) Base 4
       32, 33, ——, ——, ——, ——, ——
   (b) Base 3
       121, ——, ——, ——, ——, ——
   (c) Base 7
       64, ——, ——, ——, ——, ——
   (d) Base 5
       134, ——, ——, ——, ——, ——

6. Subtract and check your answers.
   (a)   (3 2 1 2 3) base 4   (b)   (4 3 4 4 2) base 5
        -(2 3 3 3 2) base 4   -(3 1 1 4 4) base 5
   (c)   (2 0 1 1 2) base 3   (d)   (5 4 3 4) base 6
        -(1 2 2 2 1) base 3   -(2 5 4 5) base 6
   (e)   (6 5 3 2 2 1) base 7
        -(5 4 6 6 5 3) base 7

7. Find the $(N - 1)$ complement.
   (a) (474221) base 9   (b) (654112) base 7
   (c) (431103) base 5   (d) (2301122) base 4
   (e) (115211011) base 6

8. Subtract by complements and check your results.
   (a)   (4 3 4 0 1 1 4) base 5   (b)   (6 8 4 3 0 1) base 9
        -(3 2 2 1 1 4 4) base 5   -(5 7 8 8 5 2) base 9
   (c)   (4 6 5 3 2 2) base 7   (d)   (1 2 2 0 1 1) base 3
        -(5 4 6 6 2) base 7   -(2 2 2 1) base 3
   (e)   (1 3 2 1 1) base 4   (f)   (3 5 2) base 6
        -(3 2 0 3 3) base 4   -(1 4 3 5) base 6
   (g)   (9 8 3 2) base 10
        -(1 4 3 4 5) base 10

MULTIPLICATION IN BASE *N*

As with earlier procedures, multiplication is easiest with a multiplication table. To create a multiplication table, use the following multiplications. The example shown here is for base 5.

$$0 \times 0 = 0$$
$$0 \times 1 = 0$$
$$0 \times 2 = 0 \quad\Big\} \quad 0 \times N = 0$$
$$0 \times 3 = 0$$
$$0 \times 4 = 0$$

$$1 \times 0 = 0$$
$$1 \times 1 = 1$$
$$1 \times 2 = 2 \quad\Big\} \quad 1 \times N = N$$
$$1 \times 3 = 3$$
$$1 \times 4 = 4$$

$$2 \times 0 = 0$$
$$2 \times 1 = 2$$
$$2 \times 2 = 4 \quad\Big\} \quad 2 \times N$$
$$2 \times 3 = 11 \text{ (6 is 1 five + 1 one)}$$
$$2 \times 4 = 13 \text{ (8 is 1 five + 3 ones)}$$

$$3 \times 0 = 0$$
$$3 \times 1 = 3$$
$$3 \times 2 = 11 \quad\Big\} \quad 3 \times N$$
$$3 \times 3 = 14 \text{ (9 is 1 five + 4 ones)}$$
$$3 \times 4 = 22 \text{ (12 is 2 fives + 2 ones)}$$

$$4 \times 0 = 0$$
$$4 \times 1 = 4$$
$$4 \times 2 = 13 \quad\Big\} \quad 4 \times N$$
$$4 \times 3 = 22$$
$$4 \times 4 = 31 \text{ (16 is 3 fives + 1 one)}$$

From the calculation, the multiplication table for base 5 is

×	0	1	2	3	4
0	0	0	0	0	0
1	0	1	2	3	4
2	0	2	4	11	13
3	0	3	11	14	22
4	0	4	13	22	31

Now multiply:

$$
\begin{array}{r}
2\ 1 \\
(4\ 3\ 2\ 1)\ \text{base 5} \\
\times \qquad (3)\ \text{base 5} \\
\hline
2\ 4\ 0\ 1\ 3 \leftarrow 3 \times 1 = 3 \\
\end{array}
$$

$3 \times 2 = 11$

$3 \times 3 = 14 \text{ and } 14 + 1 = 20$

$3 \times 4 = 22 \text{ and } 22 + 2 = 24$

As another example, multiply:

$$
\begin{array}{r}
(1\ 4\ 3)\ \text{base 5} \\
\times \qquad (2\ 4)\ \text{base 5} \\
\hline
1\ 2\ 3\ 2 \qquad (4 \times 143) \\
+\ 3\ 4\ 1\ 0 \qquad (20 \times 143) \\
\hline
(1\ 0\ 1\ 4\ 2)\ \text{base 5}
\end{array}
$$

This procedure can be used for any number base.

DIVISION IN BASE N

Division is the inverse operation of multiplication. Therefore, a problem
such as

$$
\begin{array}{r}
(3\ 2)\ \text{base 5} \\
\times \qquad (4)\ \text{base 5} \\
\hline
(2\ 3\ 3)\ \text{base 5}
\end{array}
$$

could be changed to either of two division problems:

$$\frac{\phantom{(4)\ \text{base}\ 5\ |}(3\ 2)\ \text{base}\ 5}{(4)\ \text{base}\ 5\ \overline{|(2\ 3\ 3)\ \text{base}\ 5}}$$

or

$$\frac{\phantom{(3\ 2)\ \text{base}\ 5\ |}(4)\ \text{base}\ 5}{(3\ 2)\ \text{base}\ 5\ \overline{|(2\ 3\ 3)\ \text{base}\ 5}}$$

Any result in division is checked by multiplication.

EXAMPLE 1

Divide:

$$(3)\ \text{base}\ 5\ \overline{|(1\ 4\ 3\ 2)\ \text{base}\ 5}}\quad (3\ 1\ 0)\quad \text{remainder}\ 2\quad \text{base}\ 5$$

$$\begin{array}{r} \underline{1\ 4}\quad (3\times3) \\ 0\ 3 \\ \underline{\ 3}\quad (3\times1) \\ 0\ 2 \\ \underline{0} \\ 2 \end{array}$$

Check by multiplying:

$$\begin{array}{r} (3\ 1\ 0)\ \text{base}\ 5 \\ \times\qquad (3)\ \text{base}\ 5 \\ \hline (1\ 4\ 3\ 0)\ \text{base}\ 5 \\ \underline{+2\quad \text{remainder}} \\ (1\ 4\ 3\ 2)\ \text{base}\ 5 \end{array}$$

Division by a two-digit number is more complicated since the multiplication table is only for the product of one-digit numbers. The method for dividing two-digit numbers involves the divisor and the set of digits for the base.

EXAMPLE 2

Divide:

$$(2\ 3)\ \text{base}\ 5\ \overline{|(4\ 2\ 4\ 2)\ \text{base}\ 5}}$$

Solution   The set of digits in base 5 is $\{0, 1, 2, 3, 4\}$. The divisor is (23) base 5.

1.   Find all products of the divisor (23) base 5 and the digits 1, 2, 3, and 4.

$$\begin{array}{cccc} (2\ 3)\ \text{base}\ 5 & (2\ 3)\ \text{base}\ 5 & (2\ 3)\ \text{base}\ 5 & (2\ 3)\ \text{base}\ 5 \\ \underline{\times\quad 1} & \underline{\times\quad 2} & \underline{\times\quad 3} & \underline{\times\quad 4} \\ 2\ 3 & 1\ 0\ 1 & 1\ 2\ 4 & 2\ 0\ 2 \end{array}$$

2. Use these products to determine the divisors. The first division would be into (42) since (4) is less than (23).

$$\begin{array}{r}
(1\ 3\ 4)\ \text{base 5} \\
(23)\ \text{base 5}\ \overline{\smash{)}(4\ 2\ 4\ 2)\ \text{base 5}} \\
\underline{2\ 3}\qquad\quad \text{(the only product} \leqslant 42) \\
1\ 4\ 4\qquad \text{(bring down the next digit)} \\
\underline{1\ 2\ 4}\qquad \text{(the only product} \leqslant 144) \\
2\ 0\ 2\qquad \text{(bring down the next digit)} \\
\underline{2\ 0\ 2}\qquad \text{(the only product} \leqslant 202) \\
0
\end{array}$$

To check:

$$\begin{array}{r}
(1\ 3\ 4)\ \text{base 5} \\
\times\qquad (2\ 3)\ \text{base 5} \\
\hline
1\ 0\ 1\ 2 \\
3\ 2\ 3\quad \\
\hline
(4\ 2\ 4\ 2)\ \text{base 5}
\end{array}$$

EXAMPLE 3

Divide:

$$(2\ 1\ 2)\ \text{base 3}\ \overline{\smash{)}(2\ 0\ 1\ 1\ 2\ 1)\ \text{base 3}}$$

Solution:

1. Find the products of 1, 2, and the divisor (2 1 2):

$$\begin{array}{r}
(2\ 1\ 2)\ \text{base 3} \\
\times\qquad 1 \\
\hline
2\ 1\ 2
\end{array}
\qquad\qquad
\begin{array}{r}
(2\ 1\ 2)\ \text{base 3} \\
\times\qquad 2 \\
\hline
1\ 2\ 0\ 1
\end{array}$$

2.

$$\begin{array}{r}
(2\ 1\ 2)\ \text{base 3} \\
(2\ 1\ 2)\ \text{base 3}\ \overline{\smash{)}(2\ 0\ 1\ 1\ 2\ 1)\ \text{base 3}} \\
\underline{1\ 2\ 0\ 1}\qquad \text{(the product} \leqslant 2011) \\
1\ 1\ 0\ 2 \\
\underline{2\ 1\ 2}\qquad \text{(the product} \leqslant 1102) \\
1\ 2\ 0\ 1 \\
\underline{1\ 2\ 0\ 1}\qquad \text{(the product} \leqslant 1201) \\
0
\end{array}$$

Check:

$$
\begin{array}{r}
(2\ 1\ 2)\ \text{base 3} \\
\times\ (2\ 1\ 2)\ \text{base 3} \\
\hline
1\ 2\ 0\ 1 \\
2\ 1\ 2 \\
1\ 2\ 0\ 1 \\
\hline
(2\ 0\ 1\ 1\ 2\ 1)\ \text{base 3}
\end{array}
$$

EXAMPLE 4

Divide:

$$(1\ 4\ 5)\ \text{base 6}\ \overline{)\,(4\ 5\ 5\ 4\ 4\ 2)\ \text{base 6}}$$

Solution   Using base 6, find the products of 1, 2, 3, 4, 5, and the divisor (1 4 5) base 6.

$$
\begin{array}{r}
(145)\ \text{base 6} \\
\times\quad 1 \\
\hline
(145)\ \text{base 6}
\end{array}
\qquad
\begin{array}{r}
(145)\ \text{base 6} \\
\times\quad 2 \\
\hline
(334)\ \text{base 6}
\end{array}
\qquad
\begin{array}{r}
(145)\ \text{base 6} \\
\times\quad 3 \\
\hline
(523)\ \text{base 6}
\end{array}
$$

$$
\begin{array}{r}
(145)\ \text{base 6} \\
\times\quad 4 \\
\hline
(1112)\ \text{base 6}
\end{array}
\qquad
\begin{array}{r}
(145)\ \text{base 6} \\
\times\quad 5 \\
\hline
(1301)\ \text{base 6}
\end{array}
$$

Divide using the products:

$$
\begin{array}{r}
(2433)\quad \text{remainder 45 base 6} \\
(145)\ \text{base 6}\ \overline{)\,(455442)\ \text{base 6}} \\
344 \\
\hline
1214 \\
1112 \\
\hline
1024 \\
523 \\
\hline
1012 \\
523 \\
\hline
45\quad \text{remainder}
\end{array}
$$

Check:

$$
\begin{array}{r}
(2433)\ \text{base 6} \\
\times\ (145)\ \text{base 6} \\
\hline
21453 \\
15020 \\
2433 \\
\hline
455353 \\
+45 \\
\hline
(455442)\ \text{base 6}
\end{array}
$$

EXERCISES

1. Multiply and fill in the base 4 table.

    $0 \times 0 =$ _____ $1 \times 0 =$ _____ $2 \times 0 =$ _____ $3 \times 0 =$ _____

    $0 \times 1 =$ _____ $1 \times 1 =$ _____ $2 \times 1 =$ _____ $3 \times 1 =$ _____

    $0 \times 2 =$ _____ $1 \times 2 =$ _____ $2 \times 2 =$ _____ $3 \times 2 =$ _____

    $0 \times 3 =$ _____ $1 \times 3 =$ _____ $2 \times 3 =$ _____ $3 \times 3 =$ _____

$\times$	0	1	2	3
0				
1				
2				
3				

2. Fill in a base 9 multiplication table.
3. Multiply.
    (a)   (1221) base 3   (b)   (2312) base 4
         $\times$ (22) base 3 $\times$ (312) base 4
    (c)   (81245) base 9   (d)   (1514) base 6
         $\times$ (76) base 9 $\times$ (244) base 6
4. Multiply and check by doing repeated additions.

$$(5714) \text{ base } 9$$
$$\times \quad (3) \text{ base } 9$$

5. Divide and check.
    (a) (2) base 3 $\overline{)(12021) \text{ base } 3}$ (b) (3) base 5 $\overline{)(240031) \text{ base } 5}$
    (c) (7) base 9 $\overline{)(1581583) \text{ base } 9}$ (d) (5) base 6 $\overline{)(431244) \text{ base } 6}$
    (e) (2) base 4 $\overline{)(13113221) \text{ base } 4}$
6. Divide and check by multiplication.
    (a) (122) base 3 $\overline{)(121020212) \text{ base } 3}$
    (b) (243) base 5 $\overline{)(10242424) \text{ base } 5}$
    (c) (33) base 4 $\overline{)(222310203) \text{ base } 4}$
    (d) (425) base 6 $\overline{)(534152) \text{ base } 6}$
    (e) (366) base 7 $\overline{)(163653123) \text{ base } 7}$
7. Divide and check the answer by repeated subtractions.

$$(4563) \text{ base } 9 \,\overline{)(20473) \text{ base } 9}$$

SUMMARY _____

Number systems can be based on other numbers besides 10. A number system using base N will have the same concepts of place value, but its place value will be determined by the power of N.

The number 10 exists in each base but has a different meaning.

(10) Base 3 means 1 three + 0 ones

or (| | |)
      1   0

(10) Base 6 means 1 six + 0 ones

or (| | | | | |)
      1   0

(10) base 10 means 1 ten + 0 ones

or (| | | | | | | | | |)
      1   0

Operations can be done in any number base, and can be simplified by using a table. The basic operations are addition and multiplication. The inverse operations are subtraction and division. Subtraction can be done using the addition of the complement. To find the complement of a base 5 number, use the $(N - 1)$ number. The base 5 number would have a 4's complement.

To subtract by complements:

$$\begin{array}{r} (43211) \text{ base 5} \\ - (31134) \text{ base 5} \\ \hline \end{array}$$

1. Find the $(N - 1)$ complement of the number to be subtracted.

    (31134) base 5   (13310) 4's complement

2. Add the complement.

$$\begin{array}{r} (43211) \text{ base 5} \\ + (13310) \quad \text{(4's complement)} \\ \hline 112021 \end{array}$$

3. Check for a carry digit.
    a. If there is a carry, add the carry to the lowest-place-value digit.

$$
\begin{array}{r}
12021 \\
+ \quad 1 \\
\hline
(12022) \text{ base } 5
\end{array}
$$

The answer is (12022) base 5.

    b. If there is no carry, the result is a negative number and it is in the $(N-1)$ complement form. To find the answer, take the complement and add a minus sign.

Division is the inverse of multiplication. It can also be done by repeated subtractions. Concepts of place value and number operations are most easily seen in a base 10 system. However, they apply to any number base and must be clearly understood to be properly applied.

CHAPTER REVIEW

1. In base 10 the place value of 3 in 7341 is _____ .
2. In base 3 the place value of 2 in (1011.102) base 3 is _____ .
3. The diagram

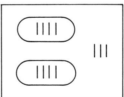

    represents a number in base _____ . The number represented is _____ _____ .
4. 10 in base 7 represents 1 _____ and 0 _____ .
5. To change (13432) base 5 to base 10:
    (a) Write the number in _____ .
    (b) _____ digits by the _____ .
    (c) _____ the products.
6. The expanded form of (1221.02) base 3 is _____ _____ .
7. The set of digits in base 4 is _____ .
8. To change a base 10 number to base N, _____ by the place values of the _____ until the _____ is zero.

SUMMARY_____

Number systems can be based on other numbers besides 10. A number system using base N will have the same concepts of place value, but its place value will be determined by the power of N.

The number 10 exists in each base but has a different meaning.

(10) Base 3 means 1 three + 0 ones

(10) Base 6 means 1 six + 0 ones

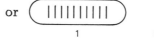

(10) base 10 means 1 ten + 0 ones

or (||||||||||)

           1   0

Operations can be done in any number base, and can be simplified by using a table. The basic operations are addition and multiplication. The inverse operations are subtraction and division. Subtraction can be done using the addition of the complement. To find the complement of a base 5 number, use the $(N-1)$ number. The base 5 number would have a 4's complement.

To subtract by complements:

$$(43211) \text{ base } 5$$
$$- (31134) \text{ base } 5$$

1. Find the $(N-1)$ complement of the number to be subtracted.

     (31134) base 5   (13310) 4's complement

2. Add the complement.

$$(43211) \text{ base } 5$$
$$+ (13310) \quad (4\text{'s complement})$$
$$\overline{112021}$$

3.   Check for a carry digit.
   a.   If there is a carry, add the carry to the lowest-place-value digit.

$$\begin{array}{r}
12021 \\
+ \quad\quad 1 \\
\hline
(12022) \text{ base } 5
\end{array}$$

   The answer is (12022) base 5.
   b.   If there is no carry, the result is a negative number and it is in the $(N-1)$ complement form. To find the answer, take the complement and add a minus sign.

Division is the inverse of multiplication. It can also be done by repeated subtractions. Concepts of place value and number operations are most easily seen in a base 10 system. However, they apply to any number base and must be clearly understood to be properly applied.

CHAPTER REVIEW_____

1. In base 10 the place value of 3 in 7341 is _____ .
2. In base 3 the place value of 2 in (1011.102) base 3 is _____ .
3. The diagram

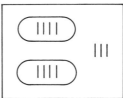

   represents a number in base _____ .   The number represented is _____
   _____ .

4. 10 in base 7 represents 1 _____ and 0 _____ .
5. To change (13432) base 5 to base 10:
   (a) Write the number in _____ .
   (b) _____ digits by the _____ .
   (c) _____ the products.
6. The expanded form of (1221.02) base 3 is _____
   _____ .

7. The set of digits in base 4 is _____ .
8. To change a base 10 number to base N, _____ by the place values of the _____ until the _____ is zero.

9. To change 76 into base 9, divide by _____. The _____ are the base 9 numbers.

10. To change a decimal in base 10 to a fractional number in base N, the method is _____ by N.

11. To change .78 to base 4: If $4 \times .78 = 3.12$, the second multiplication is _____ $\times$ _____ and the first digit of the fraction is _____.

12. A terminating decimal in base 10 is _____ a terminating decimal in base N.

13. The complement of (1323) base 4 is _____, which is the _____ complement.

14. In the subtraction by addition of complements, if there is no carry the answer is _____.

15. To check the result in subtraction, use _____.

16. The multiplication of two numbers can be done by repeated _____ _____.

17. To check the result in division, use _____.

18. Division can also be done by repeated _____.

Binary, Octal, and Hexadecimal Numbers

chapter 18

Number systems of base 2 are also known as *binary systems*. The base 2 set of digits consists of 0 and 1. The set can easily be interpreted electronically as 0 "off" and 1 "on." A base 8 system is known as an *octal system*, where 8 is a power of 2. Similarly, base 16 is called the *hexadecimal system*, where 16 is a power of 2. For this reason, binary to octal, or binary to hexadecimal, is a simple matter of interpreting groups of 0's and 1's.

BINARY NUMBERS

The binary system is one with numbers of base 2. The set of digits of base 2 is $\{0, 1\}$. The binary number has place values of powers of 2.

$$2^0 = 1 \qquad 2^{-1} = .5$$
$$2^1 = 2 \qquad 2^{-2} = .25$$
$$2^2 = 4 \qquad 2^{-3} = .125$$
$$2^3 = 8 \qquad 2^{-4} = .0625$$

$$2^4 = 16 \qquad\qquad 2^{-5} = .03125$$
$$2^5 = 32$$
$$2^6 = 64$$
$$2^7 = 128$$
$$2^8 = 256$$
$$2^9 = 512$$
$$2^{10} = 1024$$

A binary number such as

$$(101100111) \text{ binary}$$

is

$$1 \times 2^8 + 0 \times 2^7 + 1 \times 2^6 + 1 \times 2^5 + 0 \times 2^4$$
$$+ 0 \times 2^3 + 1 \times 2^2 + 1 \times 2^1 + 1 \times 2^0$$

Evaluated, it becomes

$$256 + 0 + 64 + 32 + 0 + 0 + 4 + 2 + 1 = 359$$
$$(101100111) \text{ binary} = 359 \text{ in base 10}$$

To find the binary, given the base 10 number 1257, use the following methods:

Method 1: Begin with place value and subtraction:

$$
\begin{array}{r}
1257 \\
- 1024 \\
\hline
253 \\
- 128 \\
\hline
105 \\
- 64 \\
\hline
41 \\
- 32 \\
\hline
9 \\
- 8 \\
\hline
1 \\
- 1 \\
\hline
0
\end{array}
$$

(2^{10} largest place value less than or equal to 1257)

(2^7 largest place value less than or equal to 253)

(2^6 largest place value less than or equal to 105)

(2^5 largest place value less than or equal to 41)

(2^3 largest place value less than or equal to 9)

(2^0 largest place value less than or equal to 1)

The number in binary is

$$1 \times 2^{10} + 0 \times 2^9 + 0 \times 2^8 + 1 \times 2^7 + 1 \times 2^6 + 1 \times 2^5$$
$$+ 0 \times 2^4 + 1 \times 2^3 + 0 \times 2^2 + 0 \times 2^1 + 1 \times 2^0$$

which is the number (10011101001) binary.

Method 2: Change 1257 using division:

$$
\begin{array}{rl}
2\,)\underline{1257} & \\
2\,)\underline{628} & \text{remainder } 1 \\
2\,)\underline{314} & \text{remainder } 0 \\
2\,)\underline{157} & \text{remainder } 0 \\
2\,)\underline{78} & \text{remainder } 1 \\
2\,)\underline{39} & \text{remainder } 0 \\
2\,)\underline{19} & \text{remainder } 1 \\
2\,)\underline{9} & \text{remainder } 1 \\
2\,)\underline{4} & \text{remainder } 1 \\
2\,)\underline{2} & \text{remainder } 0 \\
2\,)\underline{1} & \text{remainder } 0 \\
0 & \text{remainder } 1
\end{array}
$$

$$1257 = (10011101001) \text{ binary}$$

To change decimals involves multiplication. Change .345 to a binary number correct to 10 decimal places.

$$
\begin{array}{ll}
2 \times .345 = 0.69 & 0 \\
2 \times .69 \;\; = 1.38 & 1 \\
2 \times .38 \;\; = 0.76 & 0 \\
2 \times .76 \;\; = 1.52 & 1 \\
2 \times .52 \;\; = 1.04 & 1 \\
2 \times .04 \;\; = 0.08 & 0 \\
2 \times .08 \;\; = 0.16 & 0 \\
2 \times .16 \;\; = 0.32 & 0 \\
2 \times .32 \;\; = 0.64 & 0 \\
2 \times .64 \;\; = 1.28 & 1
\end{array}
$$

$$.345 = (.0101100001) \text{ binary}$$

The addition table in the binary number system is

+	0	1
0	0	1
1	1	10

The multiplication table in the binary system is

×	0	1
0	0	0
1	0	1

Using these two tables, do the following operations using binary numbers:

```
  111011            1101
+ 110011          ×   11
 1101110            1101
                    1101
                  100111
```

SUBTRACTION OF BINARY NUMBERS

Subtraction can be done in two ways, by the 1's complement or the 2's complement method. Using the 1's complement:

```
 1110111           1110111
- 1001101         + 0110010     (1's complement of 1001101)
                   10101001
                 +        1     (carry digit)
                    101010
```

Subtraction can also be done using 2's complements.

A *2's complement* = a 1's complement + 1

The 2's complement of 100101 is

```
the 1's complement        011010
plus 1                   +      1
                          011011
```

EXAMPLE 1

Subtract the following using 2's complements:

$$
\begin{array}{r}
1110111 \\
-\ 1001101 \\
\hline
\end{array}
\qquad
\begin{array}{r}
1110111 \\
+\ 0110011 \\
\hline
1|\,0101010 \\
\end{array}
\qquad \text{(2's complement)}
$$

$\uparrow$

drop the carry digit

EXAMPLE 2

Subtract the following using 2's complements:

$$
\begin{array}{r}
1001 \\
-\ 1111 \\
\hline
\end{array}
\qquad
\begin{array}{r}
1001 \\
+\ 0001 \\
\hline
0|\,1010 \\
\end{array}
\qquad \text{(2's complement)}
$$

$\uparrow$

if the carry digit is 0, the answer is a negative number

The answer is the 2's complement and add a negative sign. 1010 to 1's complement 0101 plus 1 is 0110 the 2's complement. So the answer is -0110.

DIVISION OF BINARY NUMBERS

Here is an example of division using binary numbers:

```
        1001000   remainder 1
  11 )11011001
      11
      ──
        00
        00
        ──
         01
         00
         ──
          11
          11
          ──
           00
           00
           ──
            00
            00
            ──
             01
             00
             ──
              1
```

The two multiplications with the digits and the divisor:

$$
\begin{array}{r}
11 \\
\times\ 0 \\
\hline
00 \\
\end{array}
\qquad
\begin{array}{r}
11 \\
\times\ 1 \\
\hline
11 \\
\end{array}
$$

To check, multiply:

$$
\begin{array}{r}
1001000 \\
\times \quad 11 \\
\hline
1001000 \\
1001000 \\
\hline
11011000 \\
+ \qquad 1 \\
\hline
11011001
\end{array}
$$

          remainder

OCTAL NUMBER SYSTEM

Octal numbers are numbers in base 8. The set of digits is $\{0, 1, 2, 3, 4, 5, 6, 7\}$. Eight is 2^3; therefore, the conversion of binary to octal is a matter of comparing three digits at a time.

Base 10	Binary	Octal
0	000	0
1	001	1
2	010	2
3	011	3
4	100	4
5	101	5
6	110	6
7	111	7

The binary number 110111001010111, separated into groups of threes is

1 1 0	1 1 1	0 0 1	0 1 0	1 1 1		binary
6	7	1	2	7		octal

To change the binary 1011011.11011 to octal:

1. Separate the number into groups of threes starting at the decimal point and going right and left. Use zeros to complete sets of threes.

two zeros added   1 zero is added

0 0 1 0 1 1 0 1 1 . 1 1 0 1 1 0

2. Change each set of three digits to the octal equivalents:

$$1 \qquad 3 \qquad 3 \qquad . \qquad 6 \qquad 6$$

Using octal-to-binary conversions involves a reversal. Each octal digit is replaced with its three-digit binary equivalent.

EXAMPLE

Change: (71042.23) octal to binary.

Solution:

7	1	0	4	2	. 2	3
111	001	000	100	010	.010	011

(71042.23) octal = (111001000100010.010011) binary

There is a natural transition between binary (base 2) and octal (base 8) numbers because $2^3 = 8$.

EXERCISES

1. Change to base 10.
   (a) (1101011) binary   (b) (111000111001) binary
   (c) (10101010.001) binary   (d) (1101.11001) binary
   (e) (111.111) binary   (f) (723) octal
   (g) (124.5) octal   (h) (5.632) octal
   (i) (70.6301) octal   (j) (0.1234) octal

2. Do the following operations in base 2.
   (a)   1101   (b)   1101101   (c) 11)1011011
       + 111   ×   101

3. Subtract using 1's complements in base 2.
   (a)   1101101   (b)   100110
       - 1001011   -   11111

   (c)   1010110   (d)   100101
       -   110001   - 110010

4. Substract using 2's complements in base 2.
   (a)   1101   (b)   11000110
       - 1010   - 10011001

   (c)   1101101   **(d)**   1000111
        –   100110   – 1101001

5. Change these base 10 numbers to (1) binary and (2) octal.
   (a) 27   **(b)** 132
   (c) 255   **(d)** .45
   (e) 372.53

6. Complete the table for binary and octal numbers.

Base 10	Binary	Octal
10	_____	_____
20	_____	_____
30	_____	_____
40	_____	_____
50	_____	_____
60	_____	_____
70	_____	_____
80	_____	_____
90	_____	_____
100	_____	_____

7. Change to binary numbers.
   (a) (47562) octal   **(b)** (1304.5) octal
   (c) (57.433) octal   **(d)** (122.16) octal
   (e) (.7542) octal

8. Change to octal numbers.
   (a) (110111) binary   **(b)** (10100.11) binary
   (c) (101.1111) binary   **(d)** (1101.01011) binary
   (e) (11.010101̄) binary

HEXADECIMAL NUMBER SYSTEM

So far all number bases were less than 10. Because of this the number sets of each base were themselves less than 10. Hexadecimal numbers are base 16. Base 16 has a number set greater than 10 since (10) base 16 represents 1 sixteen and 0 ones. Single-digit characters will have to be added to make up the number set of base 16. In working with base 16, the procedure is to use A, B, C, D, E, and F in place of two-digit numbers. The set of numbers in base 16 is

$$\{0, 1, 2, 3, 4, 5, 6, 7, 8, 9, A, B, C, D, E, F\}$$

Base 10	Hexadecimal
0	0
1	1
2	2
3	3
4	4
5	5
6	6
7	7
8	8
9	9
10	A
11	B
12	C
13	D
14	E
15	F
16	10

Place values for hexadecimals are the following:

$$16^6 = 16777216$$
$$16^5 = 1048576$$
$$16^4 = 65536$$
$$16^3 = 4096$$
$$16^2 = 256$$
$$16^1 = 16$$
$$16^0 = 1$$
$$16^{-1} = .0625$$
$$16^{-2} = .00390525$$
$$16^{-3} = .000244077625$$

EXAMPLE 1

Recognizing these place values, change (A2.9) hexadecimal to base 10.

Solution:

$$A \times 16^1 + 2 \times 16^0 + 9 \times 16^{-1} = 10 \times 16 + 2 \times 1 + 9 \times .0625$$
$$= 160 + 2 + .5625 = 162.5625$$

$$(A2.9) \text{ hexadecimal} = 162.5625$$

EXAMPLE 2

Now change 733376 base 10 to hexadecimal.

Solution:

$$
\begin{array}{rl}
16\ )\underline{733376} & \\
16\ )\underline{45836} & \text{remainder } 0 \\
16\ )\underline{2864} & \text{remainder } (12) = C \\
16\ )\underline{179} & \text{remainder } 0 \\
16\ )\underline{11} & \text{remainder } 3 \\
0 & \text{remainder } (11) = B
\end{array}
$$

733376 = (B30C0) Hexadecimal

The hexadecimal (or base 16) is a power of 2 ($2^4 = 16$). The conversion of binary to hexadecimal is a process of interpreting four binary digits at a time.

Base 10	Binary	Hexadecimal
0	0000	0
1	0001	1
2	0010	2
3	0011	3
4	0100	4
5	0101	5
6	0110	6
7	0111	7
8	1000	8
9	1001	9
10	1010	A
11	1011	B
12	1100	C
13	1101	D
14	1110	E
15	1111	F

EXAMPLE 3

Convert (110111000.101001111) binary to hexadecimal.

Solution Separate digits into groups of four starting at the decimal point and going right and left. Add zeros to complete sets of four.

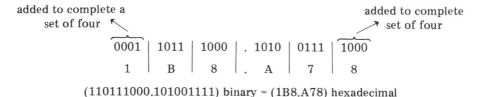

(110111000.101001111) binary = (1B8.A78) hexadecimal

EXAMPLE 4

Convert (9C.3F7) hexadecimal to binary.

Solution   Interpret each hexadecimal digit as a four-character binary number.

9	C	.	3	F	7
1001	1100	. 0011	1111	0111

(9C.3F7) hexadecimal = (10011100.001111110111) binary

HEXADECIMAL-TO-OCTAL CONVERSION

Sixteen is not a power of 8. Therefore, to convert hexadecimal to octal, the binary number must be used because 16 is a power of 2 and 8 is a power of 2. To illustrate:

$$\text{hexadecimal} \longrightarrow \text{binary} \longrightarrow \text{octal}$$

or

$$\text{octal} \longrightarrow \text{binary} \longrightarrow \text{hexadecimal}$$

EXAMPLE 1

Convert (3A2B.5) hexadecimal to octal.

Solution:

1. Hexadecimal → binary:

3	A	2	B	.	5
0011	1010	0010	1011	. 0101

2. Binary → octal:

011	101	000	101	011	. 010	100
3	5	0	5	3	. 2	4

(3A2B.5) hexadecimal = (35053.24) octal

EXAMPLE 2

Convert (1473.65) octal to hexadecimal.

Solution:

1. Octal → binary:

1	4	7	3	.	6	5
001	100	111	011	. 110	101

2. Binary → hexadecimal:

$$0011 \quad 0011 \quad 1011 \quad . \quad 1101 \quad 0100$$

$$3 \quad \quad 3 \quad \quad B \quad . \quad D \quad \quad 4$$

(1473.65) octal = (33B.D4) hexadecimal

EXERCISES

1. Complete the table for hexadecimal.

Base 10	Hexadecimal	Base 10	Hexadecimal
10	_____	100	_____
20	_____	200	_____
30	_____	300	_____
40	_____	400	_____
50	_____	500	_____
60	_____	600	_____
70	_____	700	_____
80	_____	800	_____
90	_____	900	_____

2. Fill in the missing hexadecimal numbers for each set.
    (a) 1A, 1B, 1C, ____, ____, ____, ____, ____
    (b) 23, 24, 25, 26, 27, ____, ____, ____, ____, ____
    (c) 48, 49, 4A, ____, ____, ____, ____, ____
    (d) 128, ____, ____, ____, ____, ____, ____
    (e) 2E, 2F, ____, ____, ____, ____, ____, ____

3. Change to base 10.
    (a) (8E) hexadecimal   **(b)** (200C) hexadecimal
    (c) (10A.B) hexadecimal   **(d)** (43.12A) hexadecimal
    (e) (A573.B) hexadecimal

4. Change these base 10 numbers to hexadecimal, octal, and binary.
    (a) 9756   **(b)** .5812
    (c) 187.5   **(d)** .205
    (e) 983.758

5. Change to the base specified.
    (a) (1101.111011) binary to octal
    (b) (467.55) octal to binary
    (c) (1A34.B) hexadecimal to binary
    (d) (101100111.010010001) binary to hexadecimal
    (e) (352.26) octal to hexadecimal
    (f) (4A3.DE5) hexadecimal to octal

   (g) (110.011101111) binary to base 4

   (h) (464.72) octal to base 4

   (i) (56A2.23) hexadecimal to base 4

   (j) (13211.3223) base 4 to hexadecimal

6. Complete the tables in hexadecimal.

(a)

+	0	1	2	3	4	5	6	7	8	9	A	B	C	D	E	F
0																
1																
2																
3																
4																
5																
6																
7																
8																
9																
A																
B																
C																
D																
E																
F																

(b)

×	0	1	2	3	4	5	6	7	8	9	A	B	C	D	E	F
0																
1																
2																
3																
4																
5																
6																
7																
8																
9																
A																
B																
C																
D																
E																
F																

7. Add in hexadecimal.

(a)   87A2
     + 7BA1

(b)   1B
      35
     + 48

(c)   1A4C3
     + F9E8D

8. Complete in hexadecimal.

1 + _____ = F   9 + _____ = F

2 + _____ = F   A + _____ = F

3 + _____ = F   B + _____ = F

4 + _____ = F   C + _____ = F

5 + _____ = F   D + _____ = F

6 + _____ = F   E + _____ = F

7 + _____ = F   F + _____ = F

8 + _____ = F

9. Using the results in Exercise 6, find the 15's complement of each hexadecimal number. (*Note:* F in hexadecimal is 15 in base 10.)
   (a) (87A3) hexadecimal   (b) (9AB4) hexadecimal
   (c) (1F478) hexadecimal   (d) (5F8DE) hexadecimal
   (e) (13425) hexadecimal

10. Subtract using the complements in hexadecimal.
   (a)   (873) hexadecimal   (b)   (1453) hexadecimal
       − (748) hexadecimal   − (937) hexadecimal
   (c)   (56B3) hexadecimal   (d)   (43A21) hexadecimal
       − (7AC) hexadecimal   − (58994) hexadecimal
   (e)   (2489) hexadecimal
       − (A341) hexadecimal

SUMMARY

Number bases important to the programmer are base 2, base 8, and base 16. These are commonly known as binary, octal, and hexadecimal. Binary numbers are combinations of 0's and 1's, which are easily translated into the machine interpretations "off" and "on."

Binary is converted to octal by a combination of three digits at a time, because 2^3 is 8. The hexadecimal is 2^4; a combination of four digits translates to hexadecimal.

Hexadecimal as a base 16 number needs special characters to fill in its set of digits. The special characters are:

$$10 \longrightarrow A$$
$$11 \longrightarrow B$$
$$12 \longrightarrow C$$
$$13 \longrightarrow D$$
$$14 \longrightarrow E$$
$$15 \longrightarrow F$$

CHAPTER REVIEW

1. Another name for base 2 numbers is _____ .
2. Octal numbers are numbers of base _____ .
3. Hexadecimal numbers are numbers to the base _____ .
4. The 1's complement of a binary number is formed by replacing _____ with 1's and _____ with 0's.
5. If the carry digit in complement subtraction is zero, the number is _____ .
6. A 2's complement is the _____ plus ____ .
7. The conversion from octal to binary is using ____ binary digits for each octal digit.
8. In hexadecimal the set of digits is _____ .
9. Place values in hexadecimal are powers of ____ .
10. Binary to hexadecimal is changing groups of ____ digits to a hexadecimal digit.

Computer Codes

A binary number can be interpreted in two ways. To the mathematician each digit is represented by the power of 2, according to its place value. To the computer, interpretation depends on the weighted code, which is the value assigned to each digit. These are 4-bit weighted codes, 6-bit weighted codes, and even 8-bit weighted codes.

The codes are standardized. They are determined by certain factors. These include ease of error detection, so that the transmission of electronic signals can be to some extent self-checking; simplicity of construction, for easy design and implementation in hardware; and self-complementing, for doing subtraction.

NATURAL BINARY-CODED DECIMAL

The *binary-coded decimal* or BCD code uses 4 bits. A bit represents a 0 or 1, and a weight is assigned to each bit. A natural coded binary decimal would have weights of $(8)(4)(2)(1)$. It is called natural since it has a natural power progression of the powers of 2. But it is strictly a 4-bit code, and the powers go no higher than 2^3. This means that the largest number that can be represented by 4 bits is 15. To illustrate:

$$1111 = 1 \times 8 + 1 \times 4 + 1 \times 2 + 1 \times 1 = 15$$

Base 10	Hexadecimal	BCD Code: (8)(4)(2)(1)
0	0	0000
1	1	0001
2	2	0010
3	3	0011
4	4	0100
5	5	0101
6	6	0110
7	7	0111
8	8	1000
9	9	1001
10	A	1010
11	B	1011
12	C	1100
13	D	1101
14	E	1110
15	F	1111

EXAMPLE 1

Given the BCD code

0100100101100111

Find its decimal equivalent.

Solution:

1. Separate code into 4 bits.

    0100 1001 0110 0111

2. Using the weights (8)(4)(2)(1) for each set of 4 bits from the table above, the number coded is 4967 base 10.

EXAMPLE 2

Given the same BCD number, find its base 10 value by using place values.

Solution:

$$100100101100111 = 1 \times 2^{14} + 1 \times 2^{11} + 1 \times 2^8 + 1 \times 2^6 + 1 \times 2^5$$
$$+ 1 \times 2^2 + 1 \times 2^1 + 1 \times 2^0$$
$$= 16384 + 2048 + 256 + 64 + 32 + 4 + 2 + 1$$
$$= 18759 \text{ base } 10$$

Using BCD weighted code $(8)(4)(2)(1)$, the decimal number is 4967, but using place values the number is 18759.

Example 3 illustrates another problem.

EXAMPLE 3

Change 2483 to a BCD number.

Solution   From the table: 0010 0100 1000 0011. Change 2483 to a binary number, divide by 2, and keep track of the remainders.

```
        2 )2483
        2 )1241    remainder 1      ↑
        2 )620     remainder 1
        2 )310     remainder 0
        2 )155     remainder 0
        2 )77      remainder 1
        2 )38      remainder 1
        2 )19      remainder 0
        2 )9       remainder 1
        2 )4       remainder 1
        2 )2       remainder 0
        2 )1       remainder 0
          0        remainder 1
```

$$2483 = (100110110011) \text{ binary}$$

OTHER WEIGHTED 4-BIT CODES

As mentioned earlier, codes are selected for one of the following reasons:

1. Ease of implementation on hardware
2. Simplicity of construction
3. Error detection
4. Self-complementing

The BCD code, which has weights of $(8)(4)(2)(1)$, is easy to construct because any base 10 number has a single representation.

Other possible code weights might be

$$(7)(4)(2)(1)$$

$$(5)(4)(2)(1)$$

$$(7)(4)(2)(-1)$$

$$(2)(4)(2)(1)$$

$$(8)(4)(-2)(-1)$$

etc.

The first step in setting up a code is to construct the binary representation for the set of digits in base 10.

EXAMPLE 1

Construct a binary representation for code $(7)(4)(2)(1)$.

Solution:

Base 10	Weighted code: $(7)(4)(2)(1)$
0	0000
1	0001
2	0010
3	0011
4	0100
5	0101
6	0110
7	1000
	(or 0111; if there are two representations, use the one with the fewest 1's)
8	1001
9	1010
10	1011
11	1100
12	1101
13	1110
14	1111
15	— none exists

EXAMPLE 2

Set up the weighted code $(2)(4)(2)(1)$.

Solution   The largest number that can be coded is 9 since $2 + 4 + 2 + 1 = 9$.

Base 10	Weighted code: $(2)(4)(2)(1)$
0	0000
1	0001
2	0010
3	0011
4	0100
5	1011
6	1100
7	1101
8	1110
9	1111

This code is self-complementing when forming the 9's complement. The 9's complement is N_c, such that $N + N_c = 9$. For 5 there are two choices: 1011 or 0101. In this case, rather than choosing the one with the minimum number of 1's, 1011 was chosen. The reason is that the complement of 1011 is 0100, which is 4, the 9's complement. Looking at the table, each number is seen to be self-complementing.

$$0 \longrightarrow 0000 \qquad \text{complement is } 1111 \longrightarrow 9$$
$$1 \longrightarrow 0001 \qquad \text{complement is } 1110 \longrightarrow 8$$
$$2 \longrightarrow 0010 \qquad \text{complement is } 1101 \longrightarrow 7$$

The numbers from 5 on are selected as the 9's complement.

OPERATIONS WITH WEIGHTED CODES

EXAMPLE 1

Add 4 + 7 using the weighted code (7)(4)(2)(1).

Solution:

$$
\begin{array}{r}
4 \longrightarrow 0100 \\
+ 7 \longrightarrow + 1000 \\
\hline
11 \longrightarrow 1100
\end{array}
$$

EXAMPLE 2

Subtract 13 – 5 using (7)(4)(2)(1) code and the complements.

Solution:

$$
\begin{array}{r}
13 \longrightarrow 1110 \longrightarrow 1110 \\
- 5 \longrightarrow - 0101 \longrightarrow + 1010 \\
\hline
8   ①\,1000 \\
+ 1 \\
\hline
 1001 \longrightarrow 8
\end{array}
$$

EXAMPLE 3

Subtract 57 – 98 using the weighted code (2)(4)(2)(1).

Solution:

$$
\begin{array}{r}
57 \longrightarrow 1011\ 1101 \longrightarrow 1011\ 1101 \\
- 98 \longrightarrow - 1111\ 1110 \longrightarrow + 0000\ 0001 \\
\hline
1011\ 1110
\end{array}
$$

(no carry; the answer is a negative
number in complement form)

The complement of 1011 1110 is 0100 0001, which is – 41.

EXCESS-THREE CODE

The *excess-three* (XS-3) *code* is a code formed by adding 3 to each binary-coded number.

Base 10	BCD	XS-3
0	0000	0011
1	0001	0100
2	0010	0101
3	0011	0110
4	0100	0111
5	0101	1000
6	0110	1001
7	0111	1010
8	1000	1011
9	1001	1100

In XS-3 code, no code representation consists of all zeros. It is also self-complementing.

$$3 \longrightarrow 0110 \qquad \text{complement is } 6 \longrightarrow 1001$$

EXERCISES

1. Represent 395:
   - (a) In BCD code.
   - (b) In (7)(4)(2)(1) code.
   - (c) In (2)(4)(2)(1) code.
   - (d) In XS-3 code.
   - (e) As a binary number.

2. Change 010100110110 to a base 10 number:
   - (a) Using BCD code.
   - (b) Using (7)(4)(2)(1) code.
   - (c) Using (2)(4)(2)(1) code.
   - (d) Using XS-3 code.
   - (e) As a binary number.

3. Do the addition 62 + 15:
   - (a) Using BCD code.
   - (b) Using (7)(4)(2)(1) code.
   - (c) Using (2)(4)(2)(1) code.
   - (d) Using XS-3 code.
   - (e) As a binary number.

4. Do the subtraction 75 – 32 using complements:
   - (a) In BCD code.
   - (b) In (7)(4)(2)(1) code.
   - (c) In (2)(4)(2)(1) code.
   - (d) In XS-3 code.
   - (e) As a binary number.

5. Draw a table for the base 10 digits using the weighted code (3)(3)(2)(1) such that the numbers are self-complementing for the 9's complement.

6. Draw up a table for the base 10 digits using these weighted codes.
   - (a) (5)(4)(2)(1)
   - (b) (7)(4)(2)(–1)
   - (c) (4)(4)(3)(–2)
   - (d) (8)(4)(–2)(–1)

7. Determine the weighted codes in Exercise 6 which are self-complementing.
8. Do the addition 125 + 243 using these weighted codes.
    (a) (3)(3)(2)(1)   **(b)** (5)(4)(2)(1)
    (c) (7)(4)(2)(-1)   **(d)** (4)(4)(3)(-2)
    (e) (8)(4)(-2)(-1)
9. Do the subtraction 46 - 29 using complements. Use these weighted codes.
    (a) (3)(3)(2)(1)   **(b)** (5)(4)(2)(1)
    (c) (7)(4)(2)(-1)   **(d)** (4)(4)(3)(-2)
    (e) (8)(4)(-2)(-1)
10. Develop a 4-bit weighted code for the base 10 digits 0 through 9. Use your code to:
    (a) Represent 783.
    (b) Add 3 + 4, 1 + 6, 2 + 5.
    (c) Subtract 7 - 5, 3 - 8 (use complements).

BINARY CODES OF MORE THAN 4 BITS

A 4-bit code has a maximum number of (1111) binary, which is 15 in base 10. This allows for a maximum representation of 16 characters. The computer accepts all data in binary code. Therefore, a computer code must provide for 26 alphabetic characters as well as special characters. There are more than 36 codes needed. That means that a minimum code would be a 6-bit code, since (111111) binary provides for 64 possible codes. ($2^6 = 64$). $2^5 = 32$, which is less than the required 36.

There are 6-bit, 7-bit (ASCII), and 8-bit (EBCDIC) codes. ASCII is American Standard Code for Information Interchange and EBCDIC is Extended Binary-Coded Decimal Interchange Code. The 7-bit code can provide for 2^7 or 128 character codes and the 8-bit code for 2^8 or 256 codes.

The following table shows the internal computer codes for the specified characters.

Numeric characters	6-bit	ASCII	EBCDIC
0	000 000	011 0000	1111 0000
1	000 001	011 0001	1111 0001
2	000 010	011 0010	1111 0010
3	000 011	011 0011	1111 0011
4	000 100	011 0100	1111 0100
5	000 101	011 0101	1111 0101
6	000 110	011 0110	1111 0110

Numeric characters	6-bit	ASCII	EBCDIC
7	000 111	011 0111	1111 0111
8	001 000	011 1000	1111 1000
9	001 001	011 1001	1111 1001
Alphabetic characters			
A	010 001	100 0001	1100 0001
B	010 010	100 0010	1100 0010
C	010 011	100 0011	1100 0011
D	010 100	100 0100	1100 0100
E	010 101	100 0101	1100 0101
F	010 110	100 0110	1100 0110
G	010 111	100 0111	1100 0111
H	011 000	100 1000	1100 1000
I	011 001	100 1001	1100 1001
J	100 001	100 1010	1101 0001
K	100 010	100 1011	1101 0010
L	100 011	100 1100	1101 0011
M	100 100	100 1101	1101 0100
N	100 101	100 1110	1101 0101
O	100 110	100 1111	1101 0110
P	100 111	101 0000	1101 0111
Q	101 000	101 0001	1101 1000
R	101 001	101 0010	1101 1001
S	110 010	101 0011	1110 0010
T	110 011	101 0100	1110 0011
U	110 100	101 0101	1110 0100
V	110 101	101 0110	1110 0101
W	110 110	101 0111	1110 0110
X	110 111	101 1000	1110 0111
Y	111 000	101 1001	1110 1000
Z	111 001	101 1010	1110 1001
Special characters			
Blank	110 000	010 0000	0100 0000
.	011 011	010 1110	0100 1011
(	111 100	010 1000	0100 1101
+	010 000	010 1011	0100 1110
$	101 011	010 0100	0101 1011
*	101 100	010 1010	0101 1100
)	011 100	010 1001	0101 1101
–	100 000	010 1101	0110 0000
/	110 001	010 1111	0110 0001
, (comma)	111 011	010 1100	0110 1011
=	001 011	010 1101	0111 1110

Here is how to code $1.25.

6-bit	101011 $	000001 1	011011 .	000010 2	000101 5					
Octal equivalent	101 5	011 3	000 0	001 1	011 3	011 3	000 0	010 2	000 0	101 5
7-bit	0100100 $	0110001 1	0101110 .	0110010 2	0110101 5					
8-bit	01011011 $	11110001 1	01001011 .	11110010 2	11110101 5					
Hexadecimal equivalent	0101 5	1011 B	1111 F	0001 1	0100 4	1011 B	1111 F	0010 2	1111 F	0101 5

EXERCISES

1. EBCDIC code can be written in hexadecimal. Write each EBCDIC-coded character in hexadecimal.
2. ASCII can be embedded in an 8-bit code written as ASCII-8. Find the binary for each hexadecimal code and complete the table.

Character	ASCII-8 in hexadecimal	Binary
0	50	_____
1	51	_____
2	52	_____
3	53	_____
4	54	_____
5	55	_____
6	56	_____
7	57	_____
8	58	_____
9	59	_____
A	A1	_____
B	A2	_____
C	A3	_____
D	A4	_____
E	A5	_____
F	A6	_____
G	A7	_____
H	A8	_____
I	A9	_____

Character	ASCII-8 in hexadecimal	Binary
J	AA	_____
K	AB	_____
L	AC	_____
M	AD	_____
N	AE	_____
O	AF	_____
P	B0	_____
Q	B1	_____
R	B2	_____
S	B3	_____
T	B4	_____
U	B5	_____
V	B6	_____
W	B7	_____
X	B8	_____
Y	B9	_____
Z	BA	_____
Blank	40	_____
.	4E	_____
(	48	_____
+	4B	_____
$	44	_____
*	4A	_____
)	49	_____
–	4D	_____
/	4F	_____
,	4C	_____
=	5D	_____

3. Write 07/04/76 in:
   (a) 6-bit code.   (b) 6-bit octal.
   (c) ASCII.   (d) ASCII-8
   (e) ASCII-8 hexadecimal.   (f) EBCDIC.
   (g) EBCDIC hexadecimal.

4. Given the 24-bit register contents

$$110000010100111011010100$$

   translate using:
   (a) 6-bit code.   (b) EBCDIC.
   (c) Octal.   (d) Hexadecimal.

SUMMARY _____

Coding is changing characters or base 10 numbers into 0's and 1's for interpretation by the computer. Codes can be weighted. A BCD code has weights
of (8)(4)(2)(1). Changing a number by coding is a method of assigning 4 bits
according to weights to each digit.

The 4-bit weights of (8)(4)(2)(1) are natural place values, so the code is
called a natural binary code. Other weights are chosen for certain properties.
The weighted code (2)(4)(2)(1) is a self-complementing code, for the 9's
complement of any digit is formed by reversing the 0's and 1's.

In the XS-3 code each digit has a 4-bit pattern that always includes a 1.
It also is self-complementing. A 4-bit code is limited to 2^4 or 16 characters.
The computer codes must provide for character sets that include digits, alphabetic characters, and special characters. This totals more than 36. Therefore the smallest computer code is 2^6, which provides for 64 characters.

There are 6-bit, ASCII, ASCII-8, and EBCDIC coded characters. These
are the standard computer codes and are usually specified in the programmer
operations manual.

CHAPTER REVIEW _____

1. A BCD code has _____ bits.
2. In a weighted code of (8)(4)(2)(1) the bits 1001 represent _____ .
3. A 4-bit code can represent a maximum of _____ characters.
4. BCD codes are chosen for simplicity because each digit has a _____
   _____ representation.
5. A code where 4 is 0100 and 5 is 1011 is said to be _____
   _____ .
6. The 9's complement of 3 is _____ .
7. If a weight code has two ways of representing a digit, choose the code
   with the _____ .
8. XS-3 code is BCD code plus _____ .
9. ASCII code has _____ bits.
10. The maximum character set for EBCDIC is _____ .

Mathematical Logic

Logic is a means of orderly reasoning. Logic converted into the symbols and procedures of mathematics is called mathematical logic. A basic constraint of logic is that a statement cannot be both true and false at the same time. For this reason, statements having truth value can be put together with *connectives* to form simple sentences, as well as compound sentences, and the truth value of the sentence can be determined.

A programmer uses the basic concepts of mathematical logic in writing conditionals. Most languages have the basic conditionals of mathematical logic, such as *if . . . then*, *and*, and *or*. The programmer must *interpret situations as symbols* in order to write a program to obtain the desired results.

SIMPLE SENTENCES

A statement can be symbolically interpreted as a single letter. For example, *P* can represent:

$$1 + 1 = 2$$

or

$$1 + 1 = 3$$

The basic rule of logic is that a sentence cannot be both true and false at the same time. That means if $1 + 1 = 2$ is true, $1 + 1 = 3$ cannot also be true. This property means that a specific table of truth values can be set up for any element or combination of elements.

The element P has two possibilities:

P
T
F

Two elements—P and Q—have 2^2 possible combinations.

P	Q
T	T
T	F
F	T
F	F

T's and F's were positioned in this table so as to show all possible horizontal combinations.

Three elements—P, Q, and R—have 2^3 possible combinations.

P	Q	R
T	T	T
T	T	F
T	F	T
T	F	F
F	T	T
F	T	F
F	F	T
F	F	F

Again T's and F's were positioned to show all possible horizontal combinations. The *vertical* pattern is:

First column:
    4 T's and 4 F's
Second column:
    2 T's and 2 F's, etc.
Third column:
    1 T and 1 F, etc.

CONNECTIVES

The connectives are:

$\wedge$	"and"
$\vee$	"or"
$\sim$	"not"

The "And" Sentence

"A number is between 90 and 80." This can be translated into a logical sentence by using the "and" connective.

    P: number less than 90

    Q: number greater than 80

The statement becomes: $P \wedge Q$. The possible truth combinations are:

P	Q	Possibilities	Decisions
T	T	Number is 85	True; the number is between 80 and 90.
T	F	Number is 76	False; not greater than 80.
F	T	Number is 97	False; not greater than 90.
F	F	Number is $\sqrt{-2}$	False; not a real number; it is neither greater than 80 nor less than 90.

Completing the truth table based on the decisions listed in the table above, we have

P	Q	$P \wedge Q$
T	T	T
T	F	F
F	T	F
F	F	F

The "Or" Sentence

Give a bonus to employees who have worked for 10 years or more, or are age 50 or over:

    P: an employee worked at the company for 10 or more years

    Q: an employee is 50 or older

The possible combinations are:

P	Q	Possibilities/Decisions
T	T	A *60-year-old* employee, with the company for *25 years* (bonus)
T	F	A *40-year-old* employee, with the company for *12 years* (bonus)
F	T	A *55-year-old* employee, with the company for *7 years* (bonus)
F	F	A *25-year-old* employee, with the company for *3 years* (no bonus)

The "or" condition truth table, based on the possibilities/decisions listed above, becomes

P	Q	$P \vee Q$
T	T	T
T	F	T
F	T	T
F	F	F

The "Not" Condition

(1) P	(2) $-P$	(3) $P \wedge \sim P$	(4) $P \vee \sim P$
T	F	F	T
F	T	F	T

The truth table means a statement and its negative cannot both be true at the same time. A sentence like $1 + 1 = 2$ *and* $1 + 1 \neq 2$ is always false (column 3 above). However, the "or" statement is always true: $1 + 1 = 2$ *or* $1 + 1 \neq 2$ (column 4 above).

COMPOUND SENTENCES

Combinations of simple sentences can become more complex. Combining statements with negations and/or connectives can start with a simple sentence and become a compound sentence whose truth value can be determined from a truth table. For example, start with:

P: after 4:00 P.M. of the day

Q: the off-season is July through October

R: the day is Monday through Wednesday

$\sim P$:   will be the time before 4:00 P.M. of the day

$\sim Q$:   will be one of the following months: November, December, January, February, March, April, May, June

$\sim R$:   will be any one of the following days: Thursday, Friday, Saturday, Sunday

A compound sentence might be: The rate is less if the flight is:

1.   After 4:00 P.M., or

2.   The day of the week is any day Monday through Wednesday, and is *not* in peak season.

The sentence translates into the following logical expression $P \lor (Q \land R)$.
To find the truth values for any case involves a truth table of eight entries:

	P	Q	R	$(Q \land R)$	$P \lor (Q \land R)$
1.	T	T	T	T	T
2.	T	T	F	F	T
3.	T	F	T	F	T
4.	T	F	F	F	T
5.	F	T	T	T	T
6.	F	T	F	F	F
7.	F	F	T	F	F
8.	F	F	F	F	F

The rate is less in five cases, and in three cases it is not.

If you want to make a reservation for Tuesday, March 8, at 3:00 P.M., does this qualify for a low rate?

Time is 3:00 P.M.	P is false
March 8	Q is false
Tuesday	R is true

This is line 7 of the table, and does not qualify for a lower rate.
Which of the following would qualify for a reduced rate?

a.   Thursday, August 26 at 5.00 P.M.

b.   Wednesday, September 22 at 9:00 P.M.

c.   Saturday, May 11 at 8:30 P.M.

d.   Monday, January 5 at 4:30 A.M.

e.   Tuesday, October 8 at 2:00 P.M.

The answers are:

	P	Q	R	Reduced rate?
a.	T	T	F	Yes
b.	T	F	T	Yes
c.	T	F	F	Yes
d.	F	F	T	No
e.	F	T	T	Yes

If the sentence is changed into the following:

1. The rate is less if the flight is after 4:00 P.M. or off-season, and
2. The day of the week is Monday through Wednesday.

The problem now becomes $(P \vee Q) \wedge R$. The truth table now becomes:

	P	Q	R	$(P \vee Q)$	$(P \vee Q) \wedge R$
1.	T	T	T	T	T
2.	T	T	F	T	F
3.	T	F	T	T	T
4.	T	F	F	T	F
5.	F	T	T	T	T
6.	F	T	F	T	F
7.	F	F	T	F	F
8.	F	F	F	F	F

The situation now only has three trues (qualifying conditions). Which of the following qualify for a reduced rate?

a. Thursday, August 26 at 5:00 P.M.
b. Wednesday, September 22 at 9:00 P.M.
c. Saturday, May 11 at 8:30 P.M.
d. Monday, January 5, at 4:30 A.M.
e. Tuesday, October 8 at 2:00 P.M.

The answers are:

	P	Q	R	Reduced rate?
a.	T	T	F	No
b.	T	F	T	Yes
c.	T	F	F	No
d.	F	F	T	No
e.	F	T	T	Yes

The parentheses were the only change in the two statements:

1. $P \vee (Q \wedge R)$
2. $(P \vee Q) \wedge R$

As in mathematics, parentheses make a difference.

EXERCISES

1. If a sentence has four statements, P, Q, R, and S, then:
   (a) How many possible combinations of true and false exist?
   (b) List the possibilities in a table for P, Q, R, and S.

2. If

   P is: works for the company for 5 years or more

   Q is: is a full-time employee for a minimum of 1 year

   R is: is not in a management position

   identify:
   (a) $\sim P$ (b) $\sim Q$ (c) $\sim R$

3. Draw a truth table for Exercise 2 and find the truth values for:
   (a) $P \wedge Q$ (b) $P \wedge R$
   (c) $P \vee Q$ (d) $P \vee R$
   (e) $(P \wedge Q) \vee R$ (f) $(P \vee Q) \wedge R$
   (g) $(P \wedge R) \vee (Q \wedge R)$ (h) $(P \vee R) \wedge (Q \vee R)$

4. The employees of a company qualify for a 2-week vacation if (1) they work for the company 5 years or more, or are not in a mangement position, and (2) have been full-time employees for a minimum of 1 year. In each of the following cases, determine which qualify for a 2-week vacation.
   (a) Worked 15 years for the company full time and is a department manager.
   (b) Worked for the company 10 years, 6 years part time and 4 years full time, as a stenographer in the office typing pool.
   (c) Worked for the company for 15 years as a part-time typist.
   (d) Worked 5 years full time as a machinist in the factory.
   (e) Worked 8 years part time, and for the past 6 months, full time as a receptionist.

5. Draw a truth table for

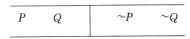

P	Q	$\sim P$	$\sim Q$

6. Using the truth table in Exercise 5, find the truth values for:
   (a) $\sim (P \wedge Q)$ (b) $\sim (P \vee Q)$
   (c) $\sim P \wedge \sim Q$ (d) $\sim P \vee \sim Q$
   (e) $P \vee \sim P$ (f) $P \wedge \sim P$
   (g) $\sim (\sim P)$

7. Using the truth table in Exercise 6, what is the negation of:
   (a) An "and" statement?   (b) An "or" statement?

THE IF—THEN CONDITIONAL

Using the statements P and Q, the conditional is: If P, then Q. The symbol for this conditional is "$\rightarrow$."

"If today is Saturday, then the school is closed."

translates logically to

P: today is Saturday
Q: school is closed

The truth table for the conditional is:

P	Q	$P \rightarrow Q$
T	T	T
T	F	F
F	T	T
F	F	T

As another example, we use

P: $x + 3 = 5$
Q: $x = 2$

If both are true, the sentence would be

"If $x + 3 = 5$, then $x = 2$."   (true statement)

But if P is true and Q is false, the sentence would be

"If $x + 3 = 5$, then x is not 2."   (false since $x = 2$)

If P is false and Q is true, the sentence would be

"If $x + 3 \neq 5$, then $x = 2$."   (true since $x = 2$ even though the problem $x + 3 \neq 5$ was incorrect)

If both are false, then:

"If $x + 3 \neq 5$, then $x \neq 2$."   (true since $x \neq 2$ if the equation
                                            $x + 3 \neq 5$)

CONVERSE

The converse is the *reversal of the conditional.*

$$Q \longrightarrow P$$

Given the statement

If $x + 3 = 5$, then $x = 2$.

The converse is

If $x = 2$, then $x + 3 = 5$.

The truth table of the converse is:

P	Q	$Q \to P$	$P \to Q$
T	T	T	T
T	F	T	F
F	T	F	T
F	F	T	T

Because the values for $P \to Q$ and $Q \to P$ are not the same, the two statements are not equivalent.

INVERSE

The inverse uses the *negatives of the conditionals:*

$$\sim P \longrightarrow \sim Q$$

P	Q	$P \to Q$	$Q \to P$	$\sim P$	$\sim Q$	$\sim P \to \sim Q$
T	T	T	T	F	F	T
T	F	F	T	F	T	T
F	T	T	F	T	F	F
F	F	T	T	T	T	T

The inverse is the equivalent of the converse, because the truth values are the same.

To illustrate, consider the statement:

"If today is December 25, then today is Christmas."

The inverse is

"If today is not December 25, then today is not Christmas."

The converse is

"If today is Christmas, then today is December 25."

CONTRAPOSITIVE

The contrapositive uses the *negatives of the converse*.

Given the converse

$$Q \longrightarrow P$$

the contrapositive is

$$\sim Q \longrightarrow \sim P$$

The truth values are:

P	Q	$P \to Q$	$\sim Q$	$\sim P$	$\sim Q \to \sim P$
T	T	T	F	F	T
T	F	F	T	F	F
F	T	T	F	T	T
F	F	T	T	T	T

The truth tables for $P \to Q$ and $\sim Q \to \sim P$ are the same, so the statements are equivalent.

Given the sentence

$$\sim P \longrightarrow Q$$

the converse is

$$Q \longrightarrow \sim P$$

The inverse is

$$\sim\sim P \longrightarrow \sim Q \qquad \text{which is} \qquad P \longrightarrow \sim Q$$

since

P	$\sim P$	$\sim(\sim P)$
T	F	T
F	T	F

Hence the statement "Two negatives make a positive!" The contrapositive is thus

$$\sim Q \longrightarrow P$$

The truth tables are:

P	Q	$\sim P$	$\sim Q$	Sentence: $\sim P \to Q$	Converse: $Q \to \sim P$	Inverse: $P \to \sim Q$	Contrapositive: $\sim Q \to P$
T	T	F	F	T	F	F	T
T	F	F	T	T	T	T	T
F	T	T	F	T	T	T	T
F	F	T	T	F	T	T	F

As another example, given:

"If you don't register, then you cannot attend class."

In symbols, this statement would be:

$$P: \quad \text{register}$$
$$Q: \quad \text{attend class}$$
$$\sim P \longrightarrow \sim Q$$

The converse is $\sim Q \to \sim P$, or

"If you cannot attend class, then you did not register."

The inverse is $\sim\sim P \rightarrow \sim\sim Q$ or $P \rightarrow Q$, or

"If you register, then you can attend class."

The contrapositive is $\sim\sim Q \rightarrow \sim\sim P$ or $Q \rightarrow P$, or

"If you can attend class, then you did register."

BICONDITIONAL

A statement such as "If P, then Q and if Q, then P" in symbolic terms is

$$(P \longrightarrow Q) \wedge (Q \longrightarrow P).$$

This is called the *biconditional*.   The biconditional can be shortened by changing the statement to

"P only if Q"

Here the truth table is:

P	Q	$(P \rightarrow Q)$	$(Q \rightarrow P)$	$(P \rightarrow Q) \wedge (Q \rightarrow P)$
T	T	T	T	T
T	F	F	T	F
F	T	T	F	F
F	F	T	T	T

The shortened form is symbolically written as

$$P \longleftrightarrow Q$$

Its truth table is:

P	Q	$P \longleftrightarrow Q$
T	T	T
T	F	F
F	T	F
F	F	T

EXERCISES

1. Write each sentence in the "if–then" form.
    (a) You are eligible for a raise after 1 year.
    (b) $3X + 5 = 14$ means that $X = 3$.
    (c) The lights will go off in the case of a power loss.
    (d) A polygon with three sides is a triangle.
    (e) To apply for a permit, one must be 16 or older.
    (f) Without a license you cannot drive a car.

2. Given the sentence, "If your sales total more than $250 per week, then your salary will include a 10% commission on sales," state:
    (a) Its converse.
    (b) Its inverse.
    (c) Its contrapositive.

3. Determine the truth value of each statement.
    (a) If $9 - 2 = 7$, then $7 + 2 = 9$.
    (b) If $4 \times -3 = -12$, then $-3/-12 = 4$.
    (c) If $8 \times 2/4 \times 4 = 1$, then $16/16 = 1$.
    (d) If $7 < 5$, then $8 < 6$.
    (e) If $2 \times 6/3 = 4$, then $45/3 \times 5 = 3$.

4. Given

    P: you are a student

    Q: you are enrolled in a mathematics class

    change each of the following logic statements into sentences.
    (a) $P \rightarrow Q$ (b) $\sim Q \rightarrow \sim P$
    (c) $Q \rightarrow P$ (d) $(\sim P \lor \sim Q)$
    (e) $\sim P \rightarrow \sim Q$ (f) $P \land \sim Q$

5. Determine the truth values of each statement.
    (a) $P \rightarrow (P \lor Q)$ (b) $P \rightarrow (P \land Q)$
    (c) $(P \land Q) \rightarrow P$ (d) $(P \lor Q) \rightarrow Q$
    (e) $(\sim P \rightarrow \sim Q) \rightarrow (P \rightarrow Q)$ (f) $\sim P \leftrightarrow (\sim P \land \sim Q)$
    (g) $(P \lor \sim Q) \leftrightarrow (P \land \sim Q)$

EQUIVALENCE

Two logical expressions are equivalent if they have the same truth value. For example:

Given: $Q \longrightarrow P$ and $\sim P \longrightarrow \sim Q$

P	Q		$\sim P$	$\sim Q$	$Q \rightarrow P$	$\sim P \rightarrow \sim Q$
T	T		F	F	T	T
T	F		F	T	T	T
F	T		T	F	F	F
F	F		T	T	T	T

truth values are the same, so
expressions are equivalent

Match the expressions that are equivalent:

1. $(P \rightarrow Q)$ A. $\sim(P \wedge \sim Q)$

2. $P \rightarrow (\sim P \wedge Q)$ B. $(P \rightarrow Q) \wedge (Q \rightarrow P)$

3. $(P \vee Q) \rightarrow (P \wedge Q)$ C. $(P \rightarrow \sim P) \wedge (P \rightarrow Q)$

To match the equivalent expressions, find the truth values for each:

(1) $(P \rightarrow Q)$	(2) $P \rightarrow (\sim P \wedge Q)$	(3) $(P \vee Q) \rightarrow (P \wedge Q)$
T	F	T
F	F	F
T	T	F
T	T	T

(A) $\sim(P \wedge \sim Q)$	(B) $(P \rightarrow Q) \wedge (Q \rightarrow P)$	(C) $(P \rightarrow \sim P) \wedge (P \rightarrow Q)$
T	T	F
F	F	F
T	F	T
T	T	T

The equivalent expressions are

1—A

2—C

3—B

TAUTOLOGIES AND CONTRADICTIONS

There are certain logical statements which are always true. An example is $P \lor \sim P$.

P	$\sim P$	$P \lor \sim P$
T	F	T
F	T	T

When the truth values are *always true*, the statement is called a *tautology*. Similarly, a logic statement that is *always false* is a *contradiction*. One such example is $P \land \sim P$.

P	$\sim P$	$P \land \sim P$
T	F	F
F	T	F

A tautology might be the statement "Today is Monday or today is not Monday." The contradiction would be the statement "Today is Monday and today is not Monday." These are obvious. But statements such as "If $a = b$, then $b = c$ if and only if $b \neq c$ and $a = b$" are not obvious. To determine if it is a tautology, a contradiction, or neither, it is necessary to draw a truth table.

$$P: \quad a = b$$
$$Q: \quad b = c$$

The logic sentence is $(P \to Q) \leftrightarrow (\sim Q \land P)$.

P	Q	$\sim Q$	$P \to Q$	$\sim Q \land P$	$(P \to Q) \leftrightarrow (\sim Q \land P)$
T	T	F	T	F	F
T	F	T	F	T	F
F	T	F	T	F	F
F	F	T	T	F	F

The statement is a contradiction since its truth value is always false. Identify the type of statement this is:

$$\text{``}x = 2 \quad \text{and} \quad y \neq 3 \quad \text{or} \quad x \neq 2 \quad \text{or} \quad y = 3\text{''}$$

$$P: \quad x = 2$$

$$Q: \quad y = 3$$

The logic sentence is $(P \land {\sim}Q) \lor ({\sim}P \lor Q)$.

P	Q	${\sim}P$	${\sim}Q$	$(P \land {\sim}Q)$	$({\sim}P \lor Q)$	$(P \land {\sim}Q) \lor ({\sim}P \lor Q)$
T	T	F	F	F	T	T
T	F	F	T	T	F	T
F	T	T	F	F	T	T
F	F	T	T	F	T	T

The statement is a tautology since it is true for each condition.

EXERCISES

Using truth tables, match the sentences that are equivalent.

1. $(P \to Q) \land (Q \to P)$ A. $(P \lor Q) \land ({\sim}(P \land Q))$
2. $P \to Q$ B. $(P \to Q) \lor {\sim}(P \leftrightarrow {\sim}Q)$
3. $P \lor {\sim}Q$ C. ${\sim}(P \to Q)$
4. $P \land {\sim}Q$ D. $(P \lor Q) \to (P \land Q)$
5. ${\sim}(P \leftrightarrow Q)$ E. ${\sim}P \to {\sim}Q$
6. $P \land (Q \lor R)$ F. $((P \land Q) \to R) \to (P \land R)$

Classify statements 7 to 16 as a tautology, a contradiction, or neither.

7. $((P \leftrightarrow Q) \land Q) \to P$ A. Tautology
8. $(P \to Q) \leftrightarrow ({\sim}P \lor Q)$ B. Contradiction
9. $(P \to Q) \lor (Q \to P)$ C. Neither
10. $((P \to Q) \land {\sim}Q) \to {\sim}P$
11. $(P \land Q \land R) \to (Q \to R)$
12. $((P \land R) \lor (P \land Q)) \to R$
13. $(P \land {\sim}Q) \lor {\sim}(P \land {\sim}Q)$
14. $(P \land {\sim}Q) \lor ({\sim}P \lor Q)$
15. $({\sim}P \lor Q) \land {\sim}({\sim}P \lor Q)$
16. ${\sim}({\sim}P \land Q) \lor {\sim}({\sim}P \lor Q)$

SUMMARY

Mathematical logic is a combination of orderly reasoning and mathematical symbols. Applying the rules means that if the truth value of individual statements is known then the truth value of the sentence can be determined.

Truth table entries depend on the number of statements involved.

Number of statements	Number of table entries
1	$2^1 = 2$
2	$2^2 = 4$
3	$2^3 = 8$
4	$2^4 = 16$
.	.
.	.
.	.
n	2^n

The basic connectives are:

$\wedge$	and
$\vee$	or
$\sim$	not
$\longrightarrow$	if, then
$\longleftrightarrow$	if and only if

Sentences formed with these connectives have truth values that depend on the connective.

P	Q	$P \wedge Q$	$P \vee Q$	$\sim P$	$P \to Q$	$P \leftrightarrow Q$
T	T	T	T	F	T	T
T	F	F	T	F	F	F
F	T	F	T	T	T	F
F	F	F	F	T	T	T

The implication statement is the if, then connective. Based on the connective,

$$\text{"if } P \text{, then } Q \text{"}$$

the converse is

$$\text{"if } Q \text{, then } P \text{"}$$

the inverse is

$$\text{``if not } P \text{, then not } Q \text{''}$$

and the contrapositive is

$$\text{``if not } Q \text{, then not } P \text{''}$$

From the truth table above, the statement and its contrapositive are equivalent. The inverse and converse are also equivalent. The truth table will also show statements that are always true or tautologies, and statements that are never true or contradictions.

CHAPTER REVIEW _____

1. The basic rule of logic is that a statement and its _____ cannot both be true.
2. A table with five statements would have _____ entries.
3. The connectives are:
   (a) $\wedge$ _____
   (b) $\vee$ _____
   (c) $\sim$ _____
   (d) $\rightarrow$ _____
   (e) $\leftrightarrow$ _____
4. For an "and" statement to be true, _____ conditions must be true.
5. For an "or" statement to be true, _____ condition must be true.
6. A statement such as "To build you must have a permit" can be written as a conditional as _____.
7. Given the statement $\sim P \rightarrow Q$:
   (a) Its inverse is _____ .
   (b) Its converse is _____ .
   (c) Its contrapositive is _____ .
8. Given $\sim P \rightarrow Q$, if P is false and Q is false, the statement is _____ .
9. The statement $(P \rightarrow Q) \wedge (Q \rightarrow P)$ is equivalent to _____ .
10. Two statements with truth tables that are the same would be _____ .
11. A statement always true is called a _____ .
12. A statement always false is called a _____ .

Boolean Algebra and Computer Logic

chapter 21 _____

Boolean algebra combines algebraic principles with mathematical logic. This allows an equation to be interpreted by a truth table or a *logic circuit*. A logic circuit, drawn from a truth table, gives a listing of all possible combinations of input pulses and the logically true output signals. This provides for the transition from logic statements to electronic circuits.

Boolean algebra is designed to simplify equations. It thus provides for simplified circuits.

The base element of a computer is the logic unit, which represents the basic "reasoning of the computer." Arithmetic operations are defined logically in a truth table. The truth table, then, becomes an algebraic expression which is simplified and becomes the logic circuit for computer operations. The circuit for an adder inputs a sequence of electrical pulses and outputs the sum.

BOOLEAN OPERATIONS

The operations in Boolean algebra are "$\cdot$" or product, and "$+$" or sum. The set of operators will be $\{0, 1\}$. The tables are:

•	0	1
0	0	0
1	0	1

+	0	1
0	0	1
1	1	1

Boolean operations have the following properties:

1. Closure: the result of each operation is always 0 or 1.
2. Identity under each operation:

> For • : $1 \bullet X = X$ The identity element is 1.
>
> For + : $0 + X = X$ The identity element is 0.

3. Commutative:

$$X + Y = Y + X$$
$$X \bullet Y = Y \bullet X$$

4. Distributive:

$$X \bullet (Y + Z) = X \bullet Y + X \bullet Z$$
$$X + (Y \bullet Z) = (X + Y) \bullet (X + Z)$$

Tables of operations can be written as truth tables using 1 and 0 to show all possible combinations of sums and products.

X	Y	$X \bullet Y$	$X + Y$
1	1	1	1
1	0	0	1
0	1	0	1
0	0	0	0

(The vertical pattern used earlier for truth tables is used here. 1 is used in place of T; 0 is used in place of F)

Here a truth table is used to verify that the identity for • is 1.

X	Y	$X \bullet Y$
1	1	1
1	0	0

Or $1 \bullet Y = Y$, since the values are the same. Besides the operations of product and sum, Boolean algebra also has a negation operation. The negation of X is written as $\overline{X}$; that is,

X	$\overline{X}$
1	0
0	1

EXAMPLE 1:

Given the sum $X + \overline{Y}$, find all possible sums.

Solution:

X	Y	$\overline{Y}$	$X + \overline{Y}$
1	1	0	1
1	0	1	1
0	1	0	0
0	0	1	1

The tables can show more than two operations.

EXAMPLE 2

Find the possible results of $X \bullet Y + Z$.

Solution   With three operators, the table has 2^3 or 8 entries:

X	Y	Z	$X \bullet Y$	$X \bullet Y + Z$
1	1	1	1	1
1	1	0	1	1
1	0	1	0	1
1	0	0	0	0
0	1	1	0	1
0	1	0	0	0
0	0	1	0	1
0	0	0	0	0

LOGIC CIRCUITS OF BOOLEAN FUNCTIONS

The basic operations in Boolean algebra are:

- $\bullet$ product
- $+$ sum
- $-$ negation

The tables are:

X	Y	X • Y	X + Y
1	1	1	1
1	0	0	1
0	1	0	1
0	0	0	0

According to mathematical logic, the • is equivalent to the "AND" and the + is equivalent to the "OR."

Operation	Boolean algebra	Mathematical logic	Logic circuit	Venn diagram
.	Product	AND		
+	Sum	OR		
−	Negation	NOT		

The circuit for X • Y is

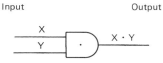

The circuit for X + Y is

The circuit for $\overline{X}$ + Y • Z is

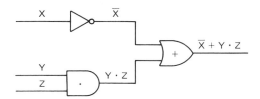

The circuit for $\overline{X} \cdot Y + X \cdot \overline{Y}$ is

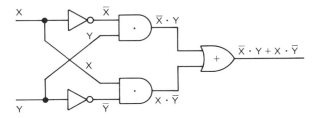

Two additional logic circuits are

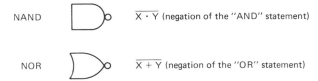

EXERCISES

1. Using truth tables, verify the distributive property.
   (a) $X \cdot (Y + Z) = X \cdot Y + X \cdot Z$
   (b) $X + (Y \cdot Z) = (X + Y) \cdot (X + Z)$

2. Find the truth table for each Boolean expression.
   (a) $\overline{X + Y}$ (b) $\overline{X \cdot Y}$
   (c) $X \cdot Y + \overline{X} \cdot \overline{Y}$ (d) $X \cdot X$
   (e) $(X + Y) \cdot \overline{(X + Y)}$

3. Determine by truth tables which of the following are true.
   (a) $X \cdot X = X$
   (b) $\overline{X} \cdot \overline{Y} = \overline{X \cdot Y}$
   (c) $\overline{X} + \overline{Y} = \overline{X + Y}$
   (d) $(X + Y) \cdot (W + Z) = X \cdot W + Y \cdot W + X \cdot Z + Y \cdot Z$
   (e) $\overline{X + Y} = \overline{X} \cdot \overline{Y}$
   (f) $\overline{X \cdot Y} = \overline{X} + \overline{Y}$
   (g) $X \cdot Y = X \cdot (\overline{X} + Y)$

4. Draw the logic circuits.
   (a) $\overline{X} + \overline{Y}$ (b) $\overline{X \cdot Y \cdot Z}$
   (c) $(X + Y) \cdot (W + Z)$ (d) $X \cdot (\overline{Y + Z})$
   (e) $\overline{X \cdot Y} + (X + Y)$

5. Find the equation and draw the truth table.
   (a)

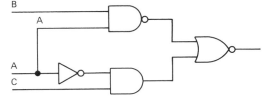

(b)

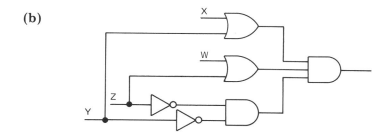

SUM-OF-PRODUCTS FORM

Given an equation, it is possible to draw a truth table. The reverse is also true. A truth table with 0's and 1's can be interpreted as an equation. The basic interpretation for the product terms is:

X is 1   then the variable is X

X is 0   then the variable is $\overline{X}$

A table with two variables is:

X	Y	Basic product terms
1	1	XY
1	0	$X\overline{Y}$
0	1	$\overline{X}Y$
0	0	$\overline{X}\,\overline{Y}$

This can be expanded for any number of terms. To find the equation from a truth table, take the sum of the product terms with truth values of 1. For example:

X	Y	F(X, Y)	Terms with truth value of 1
1	1	1 ←	XY
1	0	0	
0	1	0	
0	0	1 ←	$\overline{X}\,\overline{Y}$

The function is

$$F(X, Y) = XY + \overline{X}\,\overline{Y}$$

which is the sum of two products.

Find the function.

X	Y	Z	F(X, Y, Z)	Terms with truth value of 1
1	1	1	0	
1	1	0	1 ←	$XY\overline{Z}$
1	0	1	1 ←	$X\overline{Y}Z$
1	0	0	0	
0	1	1	0	
0	1	0	0	
0	0	1	1 ←	$\overline{X}\,\overline{Y}Z$
0	0	0	0	

The function equation is

$$F(X, Y, Z) = XY\overline{Z} + X\overline{Y}Z + \overline{X}\,\overline{Y}Z$$

Here the equation is the sum of three product terms.

PRODUCT-OF-SUMS FORM

The truth tables for the product function and sum function are:

X	Y	XY	X + Y
1	1	1	1
1	0	0	1
0	1	0	1
0	0	0	0

The product is 1 only when X is 1 and Y is 1. The reverse is true for the sum. It can be 1 in three cases, but it is 0 in only one case. The product-of-sums form is the reverse of the sum-of-products form.

X is 1   the variable is $\overline{X}$

X is 0   the variable is X

A table of values is:

X	Y	Basic sum terms
1	1	$\overline{X} + \overline{Y}$
1	0	$\overline{X} + Y$
0	1	$X + \overline{Y}$
0	0	$X + Y$

To find the equation, take the product of the sum with a truth value of 0. For example:

X	Y	F(X, Y)	Terms with truth value of 0
1	1	1	
1	0	0 ←	$\overline{X} + Y$
0	1	0 ←	$X + \overline{Y}$
0	0	1	

The equation is the product of the sums:

$$F(X, Y) = (\overline{X} + Y)(X + \overline{Y})$$

To verify this, draw the truth table for the function.

X	Y	$\overline{X}$	$\overline{Y}$	$(\overline{X} + Y)$	$(X + \overline{Y})$	$(\overline{X} + Y)(X + \overline{Y})$
1	1	0	0	1	1	1
1	0	0	1	0	1	0
0	1	1	0	1	0	0
0	0	1	1	1	1	1

The truth values of $F(X, Y)$ and $(\overline{X} + Y)(X + \overline{Y})$ are the same.

EXAMPLE 1

Find the product-of-sums equation for the following table:

X	Y	Z	F(X, Y, Z)	Terms with 0 truth value
1	1	1	0 ←	$\overline{X} + \overline{Y} + \overline{Z}$
1	1	0	1	
1	0	1	1	
1	0	0	0 ←	$\overline{X} + Y + Z$
0	1	1	0 ←	$X + \overline{Y} + \overline{Z}$
0	1	0	0 ←	$X + \overline{Y} + Z$
0	0	1	1	
0	0	0	0 ←	$X + Y + Z$

Solution   The product of the sums is

$$F(X, Y, Z) = (\overline{X} + \overline{Y} + \overline{Z})(\overline{X} + Y + Z)(X + \overline{Y} + \overline{Z})(X + \overline{Y} + Z)(X + Y + Z)$$

EXAMPLE 2

Given the function $F(X, Y) = \overline{Y}(\overline{X} + \overline{Y})$, find the sum-of-products form.

Solution:

X	Y	$\overline{X}$	$\overline{Y}$	$\overline{X} + \overline{Y}$	$\overline{Y}(\overline{X} + \overline{Y})$
1	1	0	0	0	0
1	0	0	1	1	$1 \leftarrow X\overline{Y}$
0	1	1	0	1	0
0	0	1	1	1	$1 \leftarrow \overline{X}\overline{Y}$

Use the 1 to find the sum-of-products form:

$$\overline{Y}(\overline{X} + \overline{Y}) = X\overline{Y} + \overline{X}\overline{Y}$$

EXAMPLE 3

Given the function $F(X, Y, Z) = X\overline{Y}Z + XY + XZ$, find the product-of-sums form.

Solution:

X	Y	Z	$\overline{Y}$	$X\overline{Y}Z$	XY	XZ	F(X, Y, Z)
1	1	1	0	0	1	1	1
1	1	0	0	0	1	0	1
1	0	1	1	1	0	1	1
1	0	0	1	0	0	0	0
0	1	1	0	0	0	0	0
0	1	0	0	0	0	0	0
0	0	1	1	0	0	0	0
0	0	0	1	0	0	0	0

To find the product-of-sums form, use the 0's:

$$(\overline{X} + Y + Z)(X + \overline{Y} + \overline{Z})(X + \overline{Y} + Z)(X + Y + \overline{Z})(X + Y + Z)$$

which is the product-of-sums form for

$$X\overline{Y}Z + XY + XZ$$

COMPLEMENTS OF FUNCTIONS

The complement of a function $F(X, Y)$ is defined as $\overline{F}(X, Y)$ and is formed by interchanging the 0's and 1's. Using the basic tables for $X + Y$ and $X \cdot Y$, we have

X	Y	X + Y	X • Y	$\overline{X + Y}$	$\overline{X • Y}$
1	1	1	1	0	0
1	0	1	0	0	1
0	1	1	0	0	1
0	0	0	0	1	1

The complement of $\overline{X + Y}$ resembles the product function in that it is one 1 and three 0's.

Using the table, form the equation:

X	Y	$\overline{X + Y}$	
1	1	0	
1	0	0	
0	1	0	
0	0	1	⟵ $\overline{X} • \overline{Y}$

Or $\overline{X + Y} = \overline{X} • \overline{Y}$, which implies that the complement of a sum is the product of complements.

Similarly, using the table to find the complement of $\overline{X • Y}$, we have

X	Y	$\overline{X • Y}$	
1	1	0	⟵ $\overline{X} + \overline{Y}$
1	0	1	
0	1	1	
0	0	1	

The simplest equation would be the product of sums, since there is only one 0 term. $\overline{X • Y} = \overline{X} + \overline{Y}$ or the complement of a product is the sum of the complement. Expanding this to three terms gives the following table:

X	Y	Z	X + Y + Z	$\overline{X + Y + Z}$	
1	1	1	1	0	
1	1	0	1	0	
1	0	1	1	0	
1	0	0	1	0	
0	1	1	1	0	
0	1	0	1	0	
0	0	1	1	0	
0	0	0	0	1	⟵ $\overline{X} • \overline{Y} • \overline{Z}$

Using the product of sums since there is only one 1 gives

$$\overline{X + Y + Z} = \overline{X} \cdot \overline{Y} \cdot \overline{Z}.$$

The implication is that for any number of terms, the complement of a sum is the product of complements:

$$\overline{(X + Y + Z + W + \cdots)} = \overline{X} \cdot \overline{Y} \cdot \overline{Z} \cdot \overline{W} \cdots$$

Similarly, given

$$\overline{X \cdot Y \cdot Z} = \overline{X} + \overline{Y} + \overline{Z}$$

And for any number of terms:

$$\overline{(X\,Y\,Z\,W\cdots)} = \overline{X} + \overline{Y} + \overline{Z} + \overline{W} + \cdots$$

EXAMPLE 1

Applying these two rules of complements, follow this method to find the complement of $F(X, Y) = XY + \overline{X}\overline{Y}$.

Solution:

$$\overline{F(X, Y)} = \overline{XY + \overline{X}\overline{Y}}$$

$$= \overline{XY} \cdot \overline{\overline{X}\overline{Y}} \qquad \text{the complement of a sum is the product of complements}$$

$$\overline{XY} = \overline{X} + \overline{Y}$$

$$\overline{\overline{X}\overline{Y}} = \overline{\overline{X}} + \overline{\overline{Y}} = X + Y \qquad \text{the complement of a product is the sum of complements}$$

$$\overline{F(X, Y)} = (\overline{X} + \overline{Y})(X + Y)$$

To verify this, use the following truth tables:

X	Y	X • Y	$\overline{X}$	$\overline{Y}$	$\overline{X} \cdot \overline{Y}$	$XY + \overline{X}\overline{Y}$
1	1	1	0	0	0	1
1	0	0	0	1	0	0
0	1	0	1	0	0	0
0	0	0	1	1	1	1

X	Y	X + Y	$\overline{X}$	$\overline{Y}$	$\overline{X} + \overline{Y}$	$(\overline{X} + \overline{Y})(X + Y)$
1	1	1	0	0	0	0
1	0	1	0	1	1	1
0	1	1	1	0	1	1
0	0	0	1	1	1	0

$(\overline{X} + \overline{Y})(X + Y)$ is the reverse of $XY + \overline{X}\overline{Y}$, and it is the complement of $F(X, Y)$.

EXAMPLE 2

Find the complement of $F(X, Y) = (X + \overline{Y})(\overline{X} + Y)$.

Solution   The complement is

$$\overline{F(X, Y)} = \overline{(X + \overline{Y})(\overline{X} + Y)}$$

$$= \overline{(X + \overline{Y})} \, \overline{(\overline{X} + Y)} \qquad \text{the complement of a product is the sum of complements}$$

$$\overline{(X + \overline{Y})} = \overline{X}\overline{\overline{Y}}$$

$$\overline{(\overline{X} + Y)} = \overline{\overline{X}}\overline{Y} = X\overline{Y} \qquad \text{the complement of a sum is the product of complements}$$

$$\overline{F(X, Y)} = \overline{X}Y + X\overline{Y}$$

EXAMPLE 3

Find the complement of $F(X, Y, Z) = XY\overline{Z} + X\overline{Y}Z + XYZ$.

Solution:

$$\overline{XY\overline{Z} + X\overline{Y}Z + XYZ} = \overline{(XY\overline{Z})} \bullet \overline{(X\overline{Y}Z)} \bullet \overline{(XYZ)}$$

$$\overline{XY\overline{Z}} = \overline{X} + \overline{Y} + Z$$

$$\overline{X\overline{Y}Z} = \overline{X} + Y + \overline{Z}$$

$$\overline{XYZ} = \overline{X} + \overline{Y} + \overline{Z}$$

$$\overline{F(X, Y, Z)} = (\overline{X} + \overline{Y} + Z)(\overline{X} + Y + \overline{Z})(\overline{X} + \overline{Y} + \overline{Z})$$

EXERCISES

1. Find the sum-of-products equation for the function $F(X, Y)$ having the following truth tables.

X	Y	F(X, Y)
1	1	1
1	0	1
0	1	0
0	0	0

nd the product-of-sums equation for the table in Exercise 1.

d the two equations:

  Product of sums

  Sum of products

  the following table.

X	Y	Z	F(X, Y, Z)
1	1	1	0
1	1	0	1
1	0	1	1
1	0	0	0
0	1	1	0
0	1	0	1
0	0	1	0
0	0	0	1

4. Find the complement of each equation. Verify each complement using truth tables.
   - (a) $F(X, Y) = XY + \overline{X}Y$
   - (b) $F(X, Y) = (X + \overline{Y})(\overline{X} + \overline{Y})$
   - (c) $F(X, Y, Z) = XYZ + \overline{X}Y\overline{Z} + \overline{X}YZ + \overline{X}\overline{Y}\overline{Z}$
   - (d) $F(X, Y, Z) = (X + \overline{Y} + Z)(\overline{X} + Y + Z)(X + Y + \overline{Z})(X + Y + Z)$
   - (e) $F(X, Y, Z) = (X + Y + Z)(\overline{X}\ Y\ Z)(\overline{X} + \overline{Y} + \overline{Z})(X\ \overline{Y}\ \overline{Z})$

5. Using the table

X	Y	Z	F(X, Y, Z)
1	1	1	1
1	1	0	0
1	0	1	0
1	0	0	1
0	1	1	0
0	1	0	0
0	0	1	1
0	0	0	1

find:
   - (a) The sum-of-products equation for F(X, Y, Z).
   - (b) The product-of-sums equation for F(X, Y, Z).
   - (c) The complement of the sum-of-products equation.
   - (d) The complement of the product-of-sums equation.
   - (e) The circuit logic diagram of the sum-of-products equation.
   - (f) The circuit logic diagram of the complement of the sum-of equation.

6. Change to a product of sums.
   - (a) $X + \overline{X}Y$
   - (b) $XY + \overline{Y}$
   - (c) $X + Y + Z$
   - (d) $\overline{X}Y + X\overline{Y} + \overline{X}Z + X\overline{Z}$
   - (e) $XY + XZ$

7. Change to a sum of products.
   (a) $Y(X + Y)$
   (b) $(\overline{X} + Y)(Y + \overline{Z})$
   (c) $(X + Y + Z)(\overline{X} + \overline{Y})$
   (d) $(X + Y)(X + Z)(Y + Z)$
   (e) $(X + Y + Z)(X + \overline{Y} + \overline{Z})$

KARNAUGH MAPS

The *Karnaugh map* is a pictorial representation of a function. A function with two variables has four possible combinations. Shown as a Karnaugh map, it becomes

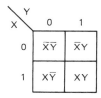

Its notation is simplified by writing

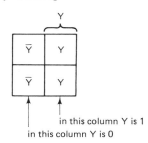

Combining the two gives

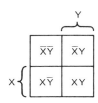

For a three-variable map, there are eight combinations:

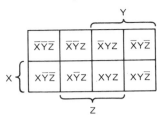

A four-variable map has 16 combinations:

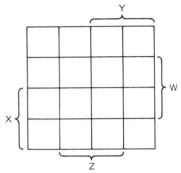

MAP OF A FUNCTION

EXAMPLE 1

Given the function

$$F(X, Y) = XY + \overline{X}Y$$

draw its Karnaugh map.

Solution   This is a two-variable function, where the map is

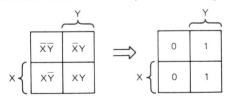

1's are used for the terms XY and $\overline{X}Y$ and 0's for all others. This is the same as the truth table since the function has 1's at XY and $\overline{X}Y$ and 0's for the other terms.

EXAMPLE 2

Given the function

$$F(X, Y, Z) = \overline{X}YZ + X\overline{Y}\overline{Z} + XY\overline{Z}$$

draw its Karnaugh map.

Solution   This is a three-variable function, and the map is

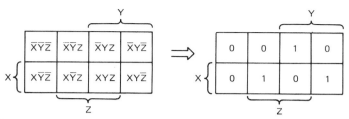

EXAMPLE 3

Given the function

$$F(W, X, Y, Z) = WXYZ + \overline{W}\overline{X}YZ + \overline{W}X\overline{Y}Z + \overline{W}X\overline{Y}\overline{Z}$$

Solution   This is a four-variable function, and the map is

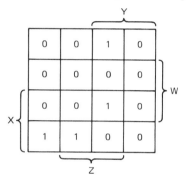

SIMPLIFYING FUNCTIONS USING KARNAUGH MAPS

In a two-variable Karnaugh map, a function = Y if the 1 is in $\overline{X}Y$ and XY.

Similarly, a function = X if 1 is in $X\overline{Y}$ and XY.

A function with adjacent 1's can be simplified by matching adjacent 1's.

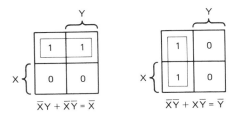

To simplify a two-variable map, circle adjacent 1's.

EXAMPLE 1

Simplify the function

$$F(X, Y) = \overline{X}Y + \overline{X}\overline{Y} + X\overline{Y}$$

Solution   The Karnaugh map is

There are two pairs of adjacent 1's, $\overline{X}$ and $\overline{Y}$. Thus $F(X, Y) = \overline{X}Y + \overline{X}\overline{Y} + X\overline{Y}$ simplifies to $F(X, Y) = \overline{X} + \overline{Y}$.

The advantage of simplifying a function is shown in this logic circuit diagram:

$$F(X, Y) = \overline{X}Y + \overline{X}\overline{Y} + X\overline{Y}$$

is

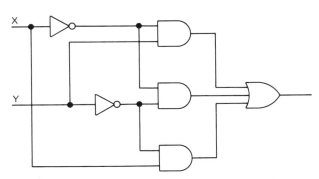

The simplified diagram having the same output given the same input signals is

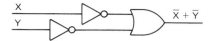

This is much easier to implement.

In a two-variable map, two adjacent 1's simplify the function. In a three-variable map, two and four adjacent 1's simplify the function.

Four adjacent 1's reduce the product from three variables to one variable:

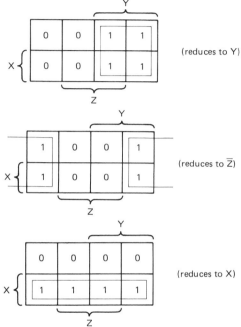

Two adjacent 1's reduce the product from three variables to two variables:

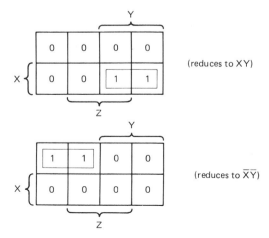

EXAMPLE 2

Find the minimal function using the following map.

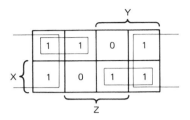

Solution   The minimal function is

$$F(X, Y, Z) = Z + XY + \overline{X}\,\overline{Y}$$

EXAMPLE 3

Graph the following function and find its minimal equation.

$$F(X, Y, Z) = \overline{X}YZ + \overline{X}\,\overline{Y}Z + \overline{X}Y\overline{Z} + XYZ + X\overline{Y}Z$$

Solution   The graph is

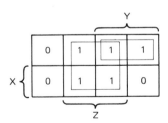

The minimal function is

$$F(X, Y, Z) = Z + \overline{X}Y$$

A four-variable function is minimized according to the following to the format:

Eight adjacent 1's reduces to a one-variable product.

Four adjacent 1's reduces to a two-variable product.

Two adjacent 1's reduces to a three-variable product.

The map is as follows:

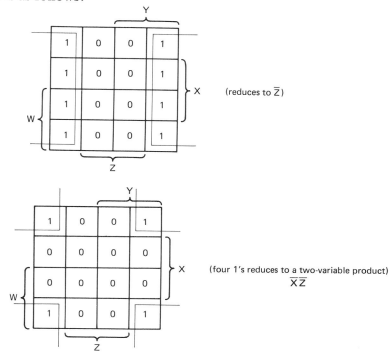

(reduces to $\overline{Z}$)

(four 1's reduces to a two-variable product)
$\overline{X}\,\overline{Z}$

EXAMPLE 4

Minimize the following function:

$$F(W, X, Y, Z) = \overline{W}X\overline{Y}\overline{Z} + \overline{W}X\overline{Y}Z + \overline{W}XYZ + \overline{W}XY\overline{Z} + WXYZ + WXY\overline{Z} + W\overline{X}\,\overline{Y}\,\overline{Z}$$
$$+ W\overline{X}\,\overline{Y}Z + W\overline{X}YZ + W\overline{X}Y\overline{Z}$$

Solution:

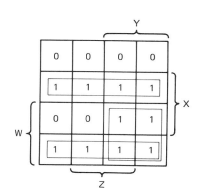

The function minimizes to

$$F(W, X, Y, Z) = \overline{W}X + W\overline{X} + WY$$

USING KARNAUGH MAPS TO FIND COMPLEMENTS

Using the function $F(X, Y, Z) = Z + XY$, find its complement and minimize. To find the complement equation, use adjacent 0's.

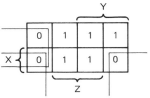

Two sets of adjacent 0's are shown in the map above. Therefore, the minimum complement of the function $F(X, Y, Z) = Z + \overline{X}Y$ is $\overline{F(X, Y, Z)} = \overline{Y}\overline{Z} + X\overline{Z}$.

EXAMPLE 1

Given the following table, find:

(a) The function
(b) The minimal function
(c) The complement

Solution:

(a) To find the function, take the sum of products with 1 in the table.

W	X	Y	Z	$F(W, X, Y, Z)$
1	1	1	1	0
1	1	1	0	0
1	1	0	1	$1 \leftarrow WX\overline{Y}Z$
1	1	0	0	$1 \leftarrow WX\overline{Y}\overline{Z}$
1	0	1	1	0
1	0	1	0	$1 \leftarrow W\overline{X}Y\overline{Z}$
1	0	0	1	$1 \leftarrow W\overline{X}\overline{Y}Z$
1	0	0	0	$1 \leftarrow W\overline{X}\overline{Y}\overline{Z}$
0	1	1	1	0
0	1	1	0	0
0	1	0	1	$1 \leftarrow \overline{W}X\overline{Y}Z$
0	1	0	0	0
0	0	1	1	0
0	0	1	0	$1 \leftarrow \overline{W}\overline{X}Y\overline{Z}$
0	0	0	1	0
0	0	0	0	$1 \leftarrow \overline{W}\overline{X}\overline{Y}\overline{Z}$

The function is

$$F(W, X, Y, Z) = WX\overline{Y}Z + WX\overline{YZ} + W\overline{X}YZ + W\overline{X}Y\overline{Z} + W\overline{XYZ}$$
$$+ \overline{W}X\overline{Y}Z + \overline{WX}Y\overline{Z} + \overline{WXYZ}$$

(b) Next, to minimize the function, draw the Karnaugh map of the function and circle the adjacent 1's.

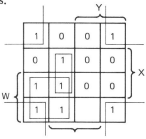

The minimal function is

$$F(W, X, Y, Z) = \overline{XZ} + W\overline{Y} + X\overline{Y}Z$$

(c) To find the complement of this function and minimize, use the same Karnaugh map, but circle adjacent 0's.

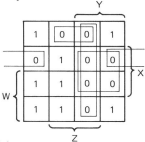

The minimal complement is

$$\overline{F(W, X, Y, Z)} = \overline{W}X\overline{Z} + \overline{WX}Z + YZ + XY$$

EXAMPLE 2

Minimize the following function:

$$F(X, Y, Z) = (\overline{X} + Y + Z)(X + \overline{Y} + Z)(X + Y + \overline{Z})(\overline{X} + \overline{Y} + Z)$$

Solution   Because $F(X, Y, Z)$ is the product of sums, the function represents the combinations that are 0's. To find the 1's, take the complement of $F(X, Y, Z)$.

$$\overline{(\overline{X} + Y + \overline{Z})(X + \overline{Y} + Z)(X + Y + \overline{Z})(\overline{X} + \overline{Y} + Z)}$$
$$= \overline{(\overline{X} + Y + \overline{Z})} + \overline{(X + \overline{Y} + Z)} + \overline{(X + Y + \overline{Z})} + \overline{(\overline{X} + \overline{Y} + Z)}$$
$$= X\overline{Y}Z + \overline{X}Y\overline{Z} + \overline{X}Y Z + XY\overline{Z}$$

The Karnaugh map is

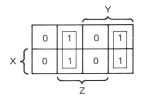

The minimal function is

$$F(X, Y, Z) = \overline{Y}Z + \overline{Z}Y$$

EXERCISES

1. Given the four-variable Karnaugh map, fill in the map showing the variable products.

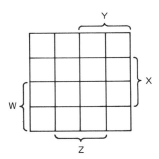

2. Draw the Karnaugh map for each function and minimize.
   (a) $F(X, Y) = XY + X\overline{Y} + \overline{X}Y$
   (b) $F(X, Y, Z) = \overline{X}\overline{Y}Z + \overline{X}YZ + X\overline{Y}Z + XYZ$
   (c) $F(X, Y, Z) = XY\overline{Z} + XYZ + X\overline{Y}Z + \overline{X}YZ + \overline{X}\overline{Y}Z$
   (d) $F(W, X, Y, Z) = WXYZ + \overline{W}XYZ + \overline{W}X\overline{Y}Z + WX\overline{Y}Z + W\overline{X}\overline{Y}Z + W\overline{X}Y\overline{Z}$
   (e) $F(W, X, Y, Z) = \overline{W}\overline{X}\overline{Y}Z + \overline{W}\overline{X}Y\overline{Z} + \overline{W}X\overline{Y}\overline{Z} + \overline{W}XY\overline{Z} + \overline{W}X\overline{Y}Z + \overline{W}XYZ$

3. Minimize.
   (a) $(\overline{X} + Z)(X + \overline{Y})(Y + \overline{Z})(X + Y + Z)$
   (b) $X\overline{Z} + Y\overline{Z} + XZ$
   (c) $XZ + \overline{X}Z + XY\overline{Z}$
   (d) $YZ + X\overline{Z} + XY + WYZ$
   (e) $(W + Y + Z)(W + Y + \overline{Z})(W + \overline{Y} + Z)(W + \overline{X})$

4. Given the tables

W	X	Y	Z	$F_1(W, X, Y, Z)$	$F_2(W, X, Y, Z)$
1	1	1	1	1	1
1	1	1	0	1	1
1	1	0	1	0	1
1	1	0	0	0	1
1	0	1	1	0	0
1	0	1	0	1	0
1	0	0	1	1	1
1	0	0	0	1	0
0	1	1	1	0	0
0	1	1	0	0	0
0	1	0	1	0	0
0	1	0	0	0	0
0	0	1	1	1	1
0	0	1	0	1	1
0	0	0	1	1	0
0	0	0	0	1	0

  (a) Find the sum-of-products form for $F_1(W, X, Y, Z)$.
  (b) Minimize the equation of F_1.
  (c) Find the minimal product-of-sums form for F_1.
  (d) Find the sum-of-products form for $F_1(W, X, Y, Z) + F_2(W, X, Y, Z)$.
  (e) Minimize $F_1 + F_2$.
  (f) Find the minimal equation of the complement of $F_1 + F_2$.
  (g) Draw the logic diagram of the minimum equation of $F_1 + F_2$.

5. Given the function

$$F(X, Y, Z) = (\overline{X} + Y + \overline{Z})(\overline{X} + \overline{Y} + \overline{Z})$$

  (a) Find the minimal sum-of-products equation.
  (b) Find the minimal product-of-sums equation.
  (c) Verify parts (a) and (b) by using a truth table.

6. Given the functions

$$F_1(W, X, Y, Z) = \overline{W}Z + XY\overline{Z} + W\overline{X}$$

$$F_2(W, X, Y, Z) = (\overline{W} + Y)(\overline{W} + X + \overline{Z})(\overline{W} + \overline{X})$$

  (a) Find the minimal sum-of-products form of $F_1 + F_2$.
  (b) Find the minimal product-of-sums form of $F_1 + F_2$.
  (c) Find the minimal sum-of-products form of $F_1 \bullet F_2$.

COMPUTER LOGIC

Computer operations are the result of input signals through the correct logic diagram. The addition of X + Y, in binary numbers, has the following table, which has two result functions, sum and carry.

X	Y	Carry	Sum
1	1	1	0
1	0	0	1
0	1	0	1
0	0	0	0

The sum equation is

$$\text{sum} = X\overline{Y} + \overline{X}Y$$

The carry equation is

$$\text{carry} = X \bullet Y$$

The logic diagram of this circuit, called the *half-adder*, is

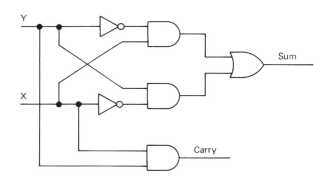

The half-adder could also be written as the product of sums.

$$\text{sum} = \overline{XY + \overline{XY}}$$ (the opposite of the zero function)
$$= (\overline{XY})\,(\overline{\overline{XY}})$$
$$= (\overline{X} + \overline{Y})\,(X + Y)$$
$$\text{carry} = \overline{X\overline{Y} + \overline{X}Y + \overline{XY}}$$

The three products in the carry equation can be simplified with a Karnaugh map:

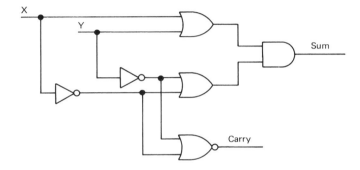

$$X\overline{Y} + \overline{X}Y + \overline{X}\,\overline{Y} = \overline{X} + \overline{Y}$$

$$\text{carry} = \overline{\overline{X} + \overline{Y}}$$

The logic circuit of the half-adder can also be drawn as

SUBTRACTION

For subtraction of X and Y, two equations are used. They are *difference* and *borrow*.

$$\begin{array}{cccc}
1 & 1 & \overset{1}{0} & 0 \\
-1 & -0 & -1 & -0 \\
\hline
0 & 1 & 1 & 0
\end{array}$$

borrow

The borrow is necessary for 0 – 1 to change the problem to 10 – 1 = 1.
The truth table for a half-subtractor is:

X	Y	Borrow	Difference
1	1	0	0
1	0	0	1
0	1	1	1
0	0	0	0

The full subtractor (subtraction truth table) will have all possible combinations of X, Y and B (borrow).

here a borrow has occurred, and changes 1 to 0

0
$\cancel{1}$
$\underline{1}$

This is the entry for the truth table:

X	Y	B
1	1	1

To solve this problem, another borrow must take place and the problem now becomes

$$\begin{array}{r} 10 \\ -\ 1 \\ \hline 1 \end{array}$$ difference is 1

This is shown as the first line of the full-subtractor truth table:

X	Y	B	Borrow	Difference
1	1	1	1	1
1	1	0	0	0
1	0	1	0	0
1	0	0	0	1
0	1	1	1	0
0	1	0	1	1
0	0	1	1	1
0	0	0	0	0

The borrow equation is

$$XYB + \overline{X}YB + \overline{X}Y\overline{B} + \overline{XY}B$$

The difference equation is

$$XYB + X\overline{Y}\overline{B} + \overline{X}Y\overline{B} + \overline{XY}B$$

EXERCISES

1. The full-adder circuit is the combination of three variables, X, Y, and C (carry from previous operation).

(a) Complete the truth table.

X	Y	C	Carry	Sum

(b) Find the sum-of-products equation for carry and sum.
(c) Draw the logic diagram of the full adder specified in part (b).
(d) Find the product-of-sums equation for carry and sum.
(e) Draw the logic diagram of the full adder specified in part (d).

2. Using the truth table for a half-subtractor given in the chapter:
   (a) Find the sum-of-product equations for borrow and difference.
   (b) Draw the logic diagram for the half-subtractor.

3. Using the truth table for a full subtractor given in the chapter:
   (a) Find the minimal sum-of-products equations for borrow and difference.
   (b) Draw the logic diagram for the full subtractor.

SUMMARY

This chapter introduced the ideas of Boolean algebra and computer logic. Using these concepts with truth tables and logic circuits, the input pulses can carry out computer operations.

Boolean operations are •, product; +, sum; and –, negation. They have the following properties:

1. Closure   3. Commutative
2. Identity   4. Distributive

The operations can be converted to logic circuits using the following symbols:

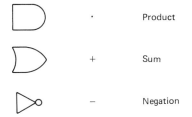

Using these symbols, any Boolean equation can be translated into a logic circuit.

There are two types of Boolean equations: "sum of products" and "product of sums." In order to implement these equations, they must usually be simplified. This is done by using Karnaugh maps. The simplification method means using adjacent 1's to minimize the variables.

The hardware application of Boolean algebra is to draw minimal circuits for computer operations.

CHAPTER REVIEW

1. The Boolean operations are ——————, ——————, and
——————.

2. Given the set {0, 1} under the + and • operations, the only results are 0 and 1. This means that the operations have ——————.

3. The identity element for sum is ——, and for the product it is ——.

4. Using the commutative property, find the result for X + Y = ——.

5. Using the distributive property, find the results for

$$X \bullet (Y + Z) = \text{——————}$$
$$X + (Y \bullet Z) = \text{——————}$$

Match the following functions to their equivalent symbols.

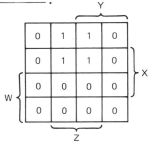

6. Product   A.

7. Sum   B.

8. Negation   C.

9. NAND   D.

10. NOR   E.

11. The following Karnaugh map of $\overline{W}\overline{X}\overline{Y}Z + \overline{W}\overline{X}YZ + \overline{W}X\overline{Y}Z + \overline{W}XYZ$ is
minimized to ——————.

Answers
to Odd-Numbered
Exercises

chapter 22 _____

Page 5

1. **(a)** $\{a, b, c, d, e, \ldots, z\}$; **(b)** $\{0, 1\}$; **(c)** $\{\ \ \}$ or ϕ ;
   (d) $\{3, 4, 5, 6, 7, 8, 9, 10, 11\}$; **(e)** $\{5, 10, 15, 20, \ldots\}$; **(f)** $\{3, 5, 7, \ldots\}$;
   (g) $\{Sept., Apr., June, Nov.\}$; **(h)** $\{Jan., Feb., Mar., May, July, Aug., Oct., Dec.\}$
3. $2^5 = 32$ subsets

$\{a, e, i, o, u\}$	$\{a, e, i\}$	$\{a, e\}$	$\{a\}$
$\{a, e, i, o\}$	$\{a, e, o\}$	$\{a, i\}$	$\{e\}$
$\{a, e, i, u\}$	$\{a, e, u\}$	$\{a, o\}$	$\{i\}$
$\{a, e, o, u\}$	$\{e, i, o\}$	$\{a, u\}$	$\{o\}$
$\{a, i. o. u\}$	$\{e, i, u\}$	$\{e, i\}$	$\{u\}$
$\{e, i, o, u\}$	$\{i, o, u\}$	$\{e, o\}$	$\{\ \ \}$
	$\{a, i, o\}$	$\{e, u\}$	
	$\{a, i, u\}$	$\{i, o\}$	
	$\{a, o, u\}$	$\{i, u\}$	
	$\{e, o, u\}$	$\{o, u\}$	

5. Number of subsets $= 2^n$

Page 6

1. (a) Finite; (b) finite; (c) infinite; (d) finite; (e) finite
3. G   5. D   7. J   9. E   11. (a) Equivalent; (b) equivalent; (c) equal;
(d) equal; (e) equal

Page 13

1. (a) {brown, orange, blue, red, white, green, yellow, black}; (b) {blue, red, white};
(c) {red, white, green, yellow}; (d) {brown, orange, green, yellow, black};
(e) {brown, orange, blue, black}; (f) {blue, red, white, green, yellow};
(g) {red, white}; (h) {brown, orange, blue, green, yellow, black};
(i) {brown, orange, black}

3. (a) 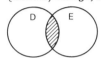   (b)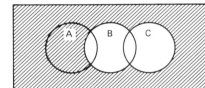

(c) 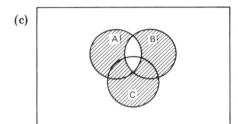   (d)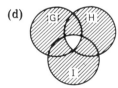

5. F   7. C   9. D

Chapter Review, Page 16

1. Collection   3. Null; 0   5. Improper   7. Infinite   9. D   11. A   13. B
15. F   17. C   19. G

CHAPTER 2

Page 19

1. (a) Rational; (b) rational; (c) irrational; (d) rational; (e) rational;
(f) rational; (g) irrational; (h) rational; (i) irrational; (j) rational

Page 20

1. (a) Integer;   (b) rational;   (c) irrational;   (d) irrational;   (e) rational;
   (f) integer;   (g) rational;   (h) rational;   (i) irrational;   (j) integer
3. (a) Rational;   (b) rational;   (c) rational;   (d) rational;   (e) rational;   (f) rational;
   (g) irrational;   (h) rational;   (i) rational;   (j) rational;   (k) neither;
   (l) irrational;   (m) rational;   (n) irrational

Page 22

1. (a) I2;   (b) I4;   (c) I4;   (d) I5;   (e) I2;   (f) I6;   (g) I3;   (h) I7;   (i) I6;
   (j) I7
3. (a) 4325;   (b) -238;   (c) -5432;   (d) _8900000;   (e) _-5400;
   (f) -114930000;   (g) ___-7134;   (h) $\overline{\text{Error}}$ 00000;   (i) Error   040603;
   (j) Error   880335

Page 25

1. (a) F4.1;   (b) F6.3;   (c) F6.5;   (d) F6.2;   (e) F11.6;   (f) F8.4;   (g) F10.8;
   (h) F8.0
3. (a) 858.0007;   (b) 858.0007394;   (c) 858.00073942;   (d) 858.000739420;
   (e) 858.0007394200;   (f) 858.000;   (g) 8.00073942;   (h) 58.0007394;
   (i) 858.000739;   (j) 858.00073;   (k) 858.0007
5. (a) 948034;   (b) 948034;   (c) 9480.34;   (d) .948034;   (e) 9.48034

Page 28

1. (a) $.782 \times 10^{-6}$;   (b) $.12 \times 10^{10}$;   (c) $.7943 \times 10^{1}$;   (d) $.5003 \times 10^{-5}$;
   (e) $.6004 \times 10^{8}$
3. (a) E7.3;   (b) E6.2;   (c) E8.4;   (d) E6.2;   (e) E6.1;   (f) E6.2;   (g) E8.4;
   (h) E6.2;   (i) E6.2;   (j) E6.2

Chapter Review, Page 29

1. Rational; irrational   3. Irrational   5. Rational   7. Irrational   9. Closed
11. Addition; subtraction; multiplication; division   13. Fixed; floating
15. seven; three   17. One more   19. $\times$ 10; exponent

CHAPTER 3

Page 33

1. (a) I6;   (b) I8;   (c) I9;   (d) I9;   (e) I6;   (f) I12;   (g) I6;   (h) I5;   (i) I12
3. (a) Format for each number: I2
        Result: I3
        Sum: 734

   (b) Number format: I2 $\times$ I2
       Result format: I3
       Result: 180
   (c) Number formats: I5, I4
       Result format: I5
       Result: 10840
   (d) Number formats:
        I2 $\times$ I2   Result: I3
        I2 $\times$ I2   Result: I3
        I1 $\times$ I2   Result: I3
       Result format: I4
       Result: 500
   (e) Number formats: I4, I4
       Result format: I5
       Result: 248

Page 40

1. (a) F9.4; (b) F6.3; (c) F8.3; (d) F8.2; (e) F10.4
3. (a) F5.1, F3.2; (b) F7.0, F3.2; (c) F5.0, F3.2; (d) F6.2, F4.3;
   (e) F7.0, F4.3
5. (a) F8.5; (b) F6.3; (c) F6.3; (d) F2.1; (e) F5.3; (f) F4.3; (g) F3.2;
   (h) F4.2; (i) F8.5; (j) F8.4

Page 44

1. $\dfrac{\text{F6.0} \times \text{F7.6}}{\text{F7.0} \times \text{F10.9}} = \dfrac{\text{F13.6}}{\text{F17.9}} = \text{F17.1}$ Remainder: F17.9

3. $1200 \times 2.55 \times (1 + .40) = \text{F5.0} \times \text{F4.2} \times (\text{F2.0} + \text{F3.2})$
                           $= \text{F9.2} \times \text{F4.2}$
                           $= \text{F 13.4}$
5. (a) I3 or F4.0; (b) I3 or F4.0; (c) I3 or F4.0; (d) F4.2; (e) F5.2;
   (f) (F4.0 – F4.0) = F4.0 or F5.0
       F4.0 $\times$ F4.2 = F8.2
       F8.2 + F5.2 = F9.2
7. (a) F3.2 $\times$ F9.2 = F12.4; (b) F3.2 $\times$ F12.4 = F15.6 $\approx$ F11.2; (c) F9.2;
   (d) F9.2 $\times$ F3.0 = F12.2; (e) F9.2 $\times$ F4.3 = F12.5 $\approx$ F9.2;
   (f) F11.2 + F12.2 + F9.2 = F13.2; (g) F3.0 $\times$ F9.2 = F12.2

Page 47

1. (a) 125; (b) 1; (c) 10001; (d) 0; (e) 0
3. (a) 98.896104%, 97.922078%, 95.454545%, 97.272727%, 64.805195%;
   (b) 99%, 98%, 95%, 97%, 65%; (c) 98.9%, 97.9%, 95.5%, 97.3%, 64.8%;
   (d) 98.90%, 97.92%, 95.45%, 97.27%, 64.81%;

(e) 1525   1523   1523
     1509   1508   1508
     1463   1471   1470
     1494   1498   1498
     1001   998   998
(f) Percent to nearest tenth

Page 49

1. 142.50   0   142.50   (a) I4; (b) F3.0; (c) F4.2; (d) F7.2; (e) F8.2;
   198.00   0   198.00   (f) F9.2
   210.00   15.75   225.75
   240.00   45.00   285.00
   168.00   0   168.00
3. (a) I3, F5.2; (b) F6.2; (c) F7.2

Chapter Review, Page 51

1. Integer   3. Decimal point   5. Whole numbers; decimal places; decimal point
7. Divisor   9. Small   11. (a) Multiply; (b) .5; (c) truncate; (d) divide

CHAPTER 4

Page 55

1. Input: 15 gallons, 19 mpg
   Process: 15 × 19 = 285 miles
   Output: 285 miles
3. Input: $1500 principal, 12% interest
   Process: 1500 × .12 × .5 = 90
           1500 + 90 = 1590
           1590 × .12 × .5 = 95.4
           90 + 95.4 = 185.40
   Output: 185.40
5. Input: 2400 square feet, $15.00 per square foot
   Process: 2400 × 15 = 36000
           36000 ÷ 12 = 3000
   Output: $3000
7. Input: $50,000 and 22%
   Process: 50000 × .22 × $\frac{30}{365}$ = $904.11
   Output: $904.11
9. Input: 6%, 135.00, 21.00, 2.50, 55.00
   Process: 135 + 21 + 2.5 + 55 = 213.55
           213.50 × .06 = 12.81
   Output: $12.81

Page 57

1. Input: 4.50, 43
   Process: Hours over 40?
           Yes: $(4.50 \times 40) + (4.50 \times 3 \times 1.5) = 200.25$
   Output: $200.25
3. Input: 26,000, 17,000, 19,000, 9500, 7000
   Process: Over 50,000?
           No: $26{,}000 \times .10 = 2600$
           No: $17{,}000 \times .10 = 1700$
           No: $19{,}000 \times .10 = 1900$
           No: $9500 \times .10 = 950$
           No: $7000 \times .10 = 700$
   Output: 2600, 1700, 1900, 950, 700
5. Input: 220.58, 210.65, 226.75, 201.90, 230.80, 12,000
   Process:
      a. $220.58 + 210.65 + 226.75 + 201.90 + 230.80 = 1090.68$
         $1090.68 \div 5 = 218.14$
      b. $12{,}000 \div 52 = 230.77$
      c. 230.77 greater than 218.14?
         Yes: 12,000 yearly wage is better.
   Output: $12,000 yearly gives a higher average weekly wage.

Page 60

1. $190,460.20   3. $25,000
5. Step 1: Product = 1
   Step 2: Counter = 1
   Step 3: Product = product $\times$ counter
   Step 4: Counter = counter + 1
   Step 5: Counter > 10?
         Yes: Output product
         No: Go to step 3
7. Step 1:   Product 1 = 1
   Step 2:   Counter = 1
   Step 3:   Product 1 = product 1 $\times$ counter
   Step 4:   Counter = counter + 1
   Step 5:   Counter > 35?
         Yes: Go to step 6
         No: Go to step 3
   Step 6:   Product 2 = 1
   Step 7:   Counter = 1
   Step 8:   Product 2 = product 2 $\times$ counter
   Step 9:   Counter = counter + 1
   Step 10: Counter > 15?
         Yes: Go to step 11
         No: Go to step 8
   Step 11: Result = product 1 − product 2
   Output: Result

Page 61

1. 2; 1; 100   3. 3; 1; 150   5. 3; 3; 123   7. 4; 6; 136   9. 9 seconds
11. Step 1: Balance = 2500, year = 20
             Rate = .075
             Month = 1
    Step 2: Balance = balance × rate + balance
    Step 3: Month = month + 1
    Step 4: Is month = year?
                Yes: Go to step 5
                No: Go to step 2
    Step 5: Output: Balance
13. Step 1: Height = 100, time = 1
    Step 2: Height = height × .375
    Step 3: Is height < = .1?
                No: Time = time + 1
                     Go to step 2
                Yes: Go to step 4
    Step 4: Output: Time

Page 62

1. 18 years   3. $4680.00   5. 1256 minutes

Page 64

1. Step 1: Counter = 0
    Step 2: Time = 12.5 A.M.   (12:30 is calculated as 12.5)
    Step 3: Bell = 1
    Step 4: Print
                Time, bells
    Step 5: Bell = bell + 1
    Step 6: Is bell > 8?
                Yes: Go to step 3
                No: Go to step 7
    Step 7: Is time > = 12.5?
                Yes: Go to step 8
                No: Time = time + .5
                     Go to step 4
    Step 8: Counter = counter + 1
    Step 9: Is counter > 2?
                Yes: stop
                No: Time = (time + .5) – 12
                     Go to step 4
3. *Machine 1*
    1   560   560   5440
    2   560   1120   4880
    3   560   1680   4320

Machine 1

4	560	2240	3760
5	560	2800	3200
6	560	3360	2640
7	560	3920	2080
8	560	4480	1520
9	560	5040	960
10	560	5600	400

Machine 2

1	1025	1025	7475
2	1025	2050	6450
3	1025	3075	5425
4	1025	4100	4400
5	1025	5125	3375
6	1025	6150	2350
7	1025	7175	1325
8	1025	8200	300

Machine 3

1	600	600	6400
2	600	1200	5800
3	600	1800	5200
4	600	2400	4600
5	600	3000	4000
6	600	3600	3400
7	600	4200	2800
8	600	4800	2200
9	600	5400	1600
10	600	6000	1000
11	600	6600	400

Machine 4

1	511.11	511.11	4288.89
2	511.11	1022.22	3777.78
3	511.11	1533.33	3266.67
4	511.11	2044.44	2755.56
5	511.11	2555.55	2244.45
6	511.11	3066.66	1733.34
7	511.11	3577.77	1222.23
8	511.11	4088.88	711.12
9	511.11	4599.99	200.01

5.	14650	1	366.25	366.25
	14300	2	357.50	723.75
	13950	3	348.75	1072.50
	⋮	⋮	⋮	⋮
	1000	40	25.00	7825.00
	650	41	16.25	7841.25
	300	42	7.50	7848.75

Chapter Review, Page 66

1. Input; process; output   3. Input   5. Output   7. Comparison   9. Loop
11. End   13. 10; .5; truncating; 10

CHAPTER 5

Page 70

1. 3; 51; 12
3. (a) Step 1: Change = cash − amount of purchase
              Output: Change
   (b) Step 1: Reduction = price × .15
       Step 2: Price = price − reduction
              Output: Price
   (c) Step 1: Percentage reached = (amount reached ÷ goal) × 100
              Output: Percentage reached
   (d) Step 1: Feet − miles × 5280
       Step 2: Inches = feet × 12
              Output: Inches
   (e) Step 1: Sale price = price − (price × .20)
       Step 2: Final amount = sale price − (sale price × .05)
              Output: Final amount

Page 74

1. (a) Step 1: Product = 1
       Step 2: Counter = 1
       Step 3: Product = product × counter
       Step 4: Counter = counter + 1
       Step 5: Counter > 8?
                   Yes: Go to "output"
                   No: Go to step 3
       Output: Product
   (b) Step 1: Sum = 0, counter = 1
       Step 2: Input: average 1
       Step 3: Sum = sum + average
       Step 4: Counter = counter + 1
       Step 5: Counter > 5?
                   Yes: Go to step 6
                   No: Go to step 2
       Step 6: Average = sum/counter
       Output: Average
   (c) Step 1: Balance = 500
              Rate = .055
              Time = 1

Step 2: Balance = balance + balance $\times$ rate
Step 3: Time = time + 1
Step 4: Time $>$ 8?
        Yes: Go to "output"
        No: Go to step 2
Output: Balance
3. Step 1: Year = 1980
        Population = 250,000
        Rate = .06
Step 2: Population = population + population $\times$ rate
Step 3: Year = year + 1
Step 4: Year $>$ 1985?
        Yes: Go to "output"
        No: Go to step 2
Output: Population
5. Step 1: Deposit = 100
        Amount = 0
        Month = 1
        Rate = .06
Step 2: Amount = amount + deposit
Step 3: Interest = amount $\times$ rate/12
Step 4: Amount = amount + interest
Step 5: Month = month + 1
Step 6: Month $>$ 60?
        Yes: Go to "output"
        No: Go to step 2
Output: Amount

Page 77

1. $247.00; $335.40; $403.75; $324.00; $508.75
3. Step 1: Investment = 500, original = 500
        Rate = .0525
        Year = 1
Step 2: Investment = investment + investment $\times$ rate
Step 3: Investment $>$ 2 $\times$ original?
        Yes: Go to "output"
        No: Go to step 4
Step 4: Year = year + 1
Step 5: Go to step 2
Output: Year
5. .37; .71; .20; .54
7. Step 1: Cost = 25,000
        Units = 0
        Save = 200
Step 2: Units = units + 2000
        Save = save + 200

Step 3: Save $>$ = cost?
            Yes: Go to "output"
            No: Go to step 2
Output: Units
9. 8824

Chapter Review, Page 78

1. Input, process, output   3. Steps   5. Sum, five   7. Yes, no

CHAPTER 6

Page 82

1. D   3. A   5. C
7. Step 1: Input: Degrees Celsius
   Step 2: Degrees Fahrenheit = $(1.8 \times N) + 32$
   Step 3: Degrees Reaumir = $(N \times 4) \div 5$
   Step 4: Degrees Kelvin = $N + 273$
   Output: Degrees Celsius, Fahrenheit, Reaumir, Kelvin
9. $176, 64, 353$

Page 87

1. C   3. A   5. C
7.

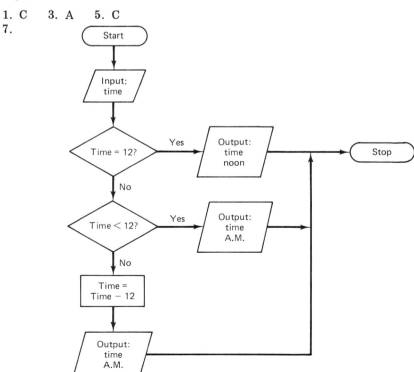

9.

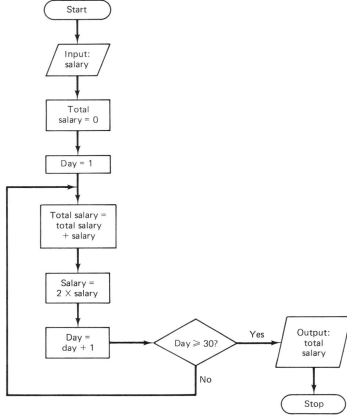

Page 92

1.

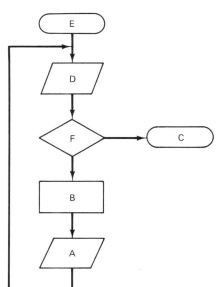

3.

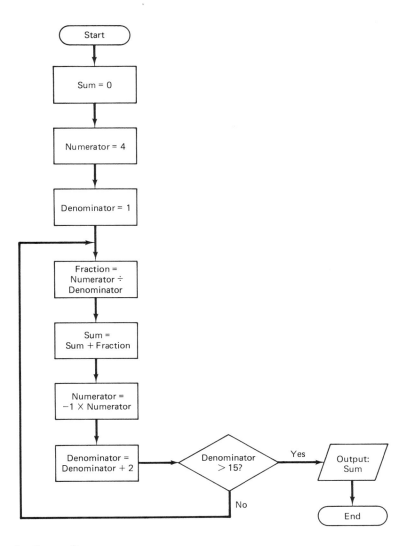

5. Step 1: Sum = 2
   Step 2: $N = 1$
   Step 3: Amount = 2
   Step 4: $N = N + 1$
   Step 5: Amount = (amount + 4) $\times$ $(-1)^N$
   Step 6: Sum = sum + amount
   Step 7: $N = 10$?
   　　　　Yes: Go to "output"
   　　　　No: Go to step 4
   Output: Sum

5. (*Continued*)

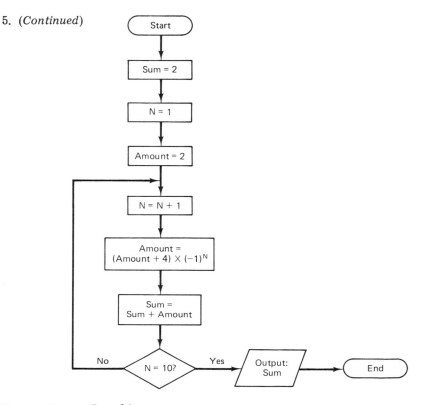

Chapter Review, Page 94

1. Flowchart   3. Process   5. Division   7. 15; 6.5   9. Price 1 > price 2

CHAPTER 7

Page 99

1.
	Variables	*Constants*
(a)	a; b	4
(b)	c, b, d	None
(c)	e	4, 11
(d)	f	36
(e)	I, Y	12, 3

3. (a) $I = 36Y$; (b) $S = C + P$; (c) $D = S \times T$; (d) $D = M \div V$; (e) $I = R \times T \times P$;
   (f) $P = 2(L + W)$; (g) $0 = 1.5 \times R \times (H - 40)$

5. (a) /, +, 11.5; (b) *, /, *, 18; (c) *, +, 14; (d) /, +, 13.4;
   (e) /, *, /, 12; (f) +, −, +, 11; (g) −, /, *, 4; (h) *, +, *, *, 72;
   (i) −, *, +, *, 85; (j) /, *, /, 4; (k) /, /, /, .25

Page 102

1. (a) $3 * X - 5 * Y$;   (b) $5 * A + C - B$;   (c) $A + B * (D - 2)$;
   (d) $(4 * A - 3 * B)/(X + 2 * Y)$;   (e) $A/5 * C + 32$;   (f) $5/9 * (F - 32)$;
   (g) $6 - (X + 2)/5$;   (h) $X/(X + 2) - 2/(X - 2)$;   (i) $(A + B + C)/2$;
   (j) $B * * 2 - 4 * A * C$;   (k) $S * (S - A) * (S - B) * (S - C)$

Page 106

1. (a) $C D +$;   (b) $F D F * +$;   (c) $A G D - +$;   (d) $D C B A + + *$;
   (e) $A B C * + A C * /$
3. $XY * 5Y * - YWX * + /$; 1

Chapter Review, Page 107

1. Equals; is assigned to   3. Constant
5. (a) $+$;   (b) $-$;   (c) $*$;   (d) $/$;   (e) $* *$
7. Parentheses   9. Parentheses are not needed

CHAPTER 8

Page 111

1. (a) $8 < 15$;   (b) $X \leqslant 10$;   (c) $D \leqslant 15$;   (d) $A \geqslant 16$;   (e) $D > 1000$
3. (a) $X > 2$;   (b) $X \leqslant -3$;   (c) $X < 9$;   (d) $X \geqslant -6$

Page 114

1. (a) $7, 9$;   (b) $0, 1, 2, 3$;   (c) $-5, 10$;   (d) $6, 8, 10$
3. (a)

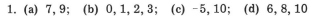

(b)

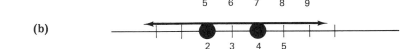

(c)

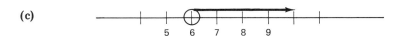

(d)

(e)

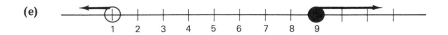

Chapter Review, Page 116

1. Inequality   3. $<$; $>$ 5. $\leqslant$ 7. Reversed   9. Either is true

CHAPTER 9

Page 119

1. (a) 7^8;   (b) C^4;   (c) A^2B^5;   (d) $4^4A^2D^4$
3. (a) $\frac{1}{5}$, 0.2;   (b) $\frac{1}{2}$, 0.5;   (c) $\sqrt[8]{}$, 0.125;   (d) $\sqrt[3]{}$, 0.333;   (e) $\sqrt[8]{}$, $\frac{1}{8}$;
   (f) $\sqrt[4]{}$, $\frac{1}{4}$;   (g) $\sqrt{}$, $\frac{1}{2}$;   (h) $(()^{1/2})^{1/5}$, $(()^{.5})^{.2}$;   (i) $\sqrt{\sqrt{}}$, $(()^{.5})^{.25}$;
   (j) $\sqrt{\sqrt{}}$, $(()^{1/2})^{1/4}$
5. (a) 2.87102;   (b) 2.98685;   (c) 2.99868;   (d) 2.99987;   (e) 2.99999;   (f) 3.;
   (g) 3.06665;   (h) 3.0066;   (i) 3.00066;   (j) 3.00007;   (k) 3.00001;   (l) 3

Page 122

1. (a) $(P)^{10}$;   (b) $(R)^{12}$;   (c) $(S)^3$;   (d) $(T)^{-11}$;   (e) A^{10}/D^{12};   (f) $X^8Y^6W^{10}$;
   (g) AB^{-2};   (h) X^2;   (i) $9C^2D^4$;   (j) $3^2 5AB^2D^2$
3. $(27*X**3*Y**5)**.5/(3*X*Y**2)**.5$;   (a) 10;
   (b) $\sqrt{9X^2Y^3} = (9*X**2*Y**3)**.5$;   (c) 5

Chapter Review, Page 123

1. Base; exponent   3. Square root   5. Root   7. $\frac{1}{2}$ 9. More accurate
11. Fourth; A

CHAPTER 10

Page 128

1. (a) X = D − C;   (b) X = A + 5;   (c) X = 8Y;   (d) X = F/7;   (e) X = G/4;
   (f) X = Y − 17;   (g) X = 11 + D;   (h) X = 5P;   (i) X = AB/9;   (j) X = C/A
3. (a) 56;   (b) 126;   (c) 4;   (d) 115;   (e) 16;   (f) 24

Page 131

1. (a) A/W;   (b) 2A/B;   (c) $\sqrt{V/H}$;   (d) A/H − B;   (e) AN − K + L;
   (f) $(I+1)^N * P$;   (g) M^2/T;   (h) $\sqrt{C^2 - B^2}$;   (i) $(V/125.3)^6/Z$;   (j) $(T/2\pi)^2 * 32.2$

Page 134

1. 12   3. 6.4   5. 72,000   7. 14,960   9. second model   11. 2414027.80
13. (a) 57;   (b) 21

Chapter Review, Page 135

1. A * B; C   3. Inverse   5. Subtracting 11   7. AB/C

CHAPTER 11

Page 139

1. (b); (c); (d); (f); (g); (h); (i)
3. (a) $-7, -5$; (b) $-9, -8$; (c) $-8, 3$; (d) $71, -2$; (e) $12, 11$; (f) $-9, -9$;
   (g) $14, 14$; (h) $-\frac{5}{3}, 12$; (i) $\frac{9}{2}, \frac{17}{5}$; (j) $-\frac{3}{7}, -\frac{5}{8}$

Page 142

1. (a) $x^2 - 11x + 24$; (b) $x^2 + 17x + 60$; (c) $x^2 + 5x - 84$; (d) $x^2 - 2x - 3$;
   (e) $x^2 + 0x - 729$
3. (a) $5, 2$; (b) $9, 7$; (c) $18, 2$; (d) $9, 4$; (e) $5, 8$

Page 152

1. (a) $-8, -4$; (b) $\frac{3}{5}, -1$; (c) $\frac{1}{3}, \frac{9}{2}$; (d) $-\frac{2}{7}, 1$; (e) $\frac{2}{3}, -\frac{4}{3}$
3. (a) $6, 2$; (b) $-\frac{5}{3}, -\frac{1}{2}$; (c) Does not exist; (d) $\dfrac{9 + \sqrt{129}}{8}, \dfrac{9 - \sqrt{129}}{8}$;
   (e) $\dfrac{6 + \sqrt{22}}{7}, \dfrac{6 - \sqrt{22}}{7}$; (f) $-\frac{11}{5}, -1$; (g) $\frac{7}{2}, \frac{7}{2}$; (h) $\dfrac{-9 + \sqrt{185}}{4}$,
   $\dfrac{-9 - \sqrt{185}}{4}$; (i) $\frac{19}{11}, -7$; (j) $\dfrac{10 + \sqrt{145}}{9}, \dfrac{10 - \sqrt{145}}{9}$
5. (a) $(\frac{5}{2}, -\frac{35}{4})$; (b) $(\frac{3}{2}, -\frac{3}{2})$; (c) $(\frac{11}{10}, \frac{299}{20})$; (d) $(2, 24)$; (e) $(0, -16)$;
   (f) $(0, -30)$; (g) $(-\frac{5}{6}, -\frac{25}{12})$; (h) $(\frac{3}{16}, \frac{55}{32})$; (i) $(\frac{1}{8}, -\frac{49}{16})$; (j) $(0, -25)$

Chapter Review, Page 154

1. $Ax^2 + Bx + C$ 3. 0   5. Sum; product   7. $\dfrac{-B + \sqrt{B^2 - 4AC}}{2A}$
9. (a) 2; (b) 1; (c) none   11. 2; rational   13. Does not

CHAPTER 12

Page 159

1. (a) $7x - 3y = -8$; (b) $2x - 4y = -11$; (c) $3x - 5y = 10$; (d) $8x - 7y = 6$;
   (e) $3x - 2y = 12$; (f) $9x - 2y = 12$; (g) $4x - 3y = 11$; (h) $7x + 3y = 20$;
   (i) $3x - 4y = -6$; (j) $15x - 10y = 3$

3. D   5. A; B; D   7. I   9. O   11. P

13. **(a)** $(0, 5), (-3, 0)$ **(b)** $(0, 3), (-2, 0)$

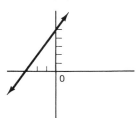   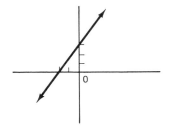

(c) $(0, 14), (4, 0)$ **(d)** $(0, \frac{7}{2}), (\frac{7}{3}, 0)$

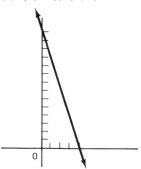   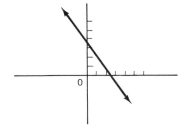

(e) $(0, -\frac{7}{4}), (-\frac{7}{5}, 0)$

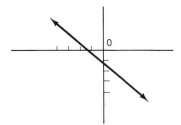

Page 165

1. **(a)** $y = \frac{5}{3}x + \frac{7}{3}, \frac{5}{3}, (0, \frac{7}{3})$;   **(b)** $y = -4x + 6, -4, (0, 6)$;
   (c) $y = -\frac{3}{2}x + \frac{15}{2}, -\frac{3}{2}, (0, \frac{15}{2})$;   **(d)** $y = -7x + 14, -7, (0, 14)$;
   (e) $y = \frac{11}{2}x - 22, \frac{11}{2}, (0, -22)$;   **(f)** $y = -\frac{8}{3}x - 8, -\frac{8}{3}, (0, -8)$;
   (g) $y = \frac{4}{7}x + 4, \frac{4}{7}, (0, 4)$;   **(h)** $y = -\frac{1}{6}x + 4, -\frac{1}{6}, (0, 4)$;
   (i) $y = \frac{1}{3}x - 3, \frac{1}{3}, (0, -3)$;   **(j)** $y = -\frac{5}{7}x + 5, -\frac{5}{7}, (0, 5)$

3. (a) $2x + 3y = 5$;   (b) $x + y = 8$;   (c) $5x - 6y = 23$;   (d) $2x - 5y = -9$;
   (e) $7x - 4y = -17$;   (f) $4x + 7y = 30$;   (g) $5x + 7y = -11$;   (h) $3x - 2y = -1$;
   (i) $9x + y = 19$;   (j) $3x - 11y = 2$;
5. (a) $y = -\frac{2}{3} x + \frac{5}{3}$;   (b) $y = -x + 8$;   (c) $y = \frac{5}{6} x - \frac{23}{6}$;   (d) $y = \frac{2}{5} x + \frac{9}{5}$;
   (e) $y = \frac{7}{4} x + \frac{17}{4}$;   (f) $y = -\frac{4}{7} x + \frac{30}{7}$;   (g) $y = -\frac{5}{7} x - \frac{11}{7}$;   (h) $y = \frac{3}{2} x + \frac{1}{2}$;
   (i) $y = -9x + 19$;   (j) $y = \frac{3}{11} x - \frac{2}{11}$

Page 169

1. (a) $(-1, 2)$;   (b) $(2, 6)$;   (c) $(-6, 5)$;   (d) $(4, -4)$;   (e) No solution;   (f) $(9, -2)$;
   (g) $(-\frac{42}{13}, \frac{89}{13})$;   (h) $(-3, 0)$;   (i) No solution;   (j) No solution
3. H   5. A   7. B   9. D
11. (a) $B * D - A * E$;   (b) $X = -(B * F - C * E)/(B * D - A * E)$;
    (c) $Y = -(C * D - A * F)/(B * D - A * E)$

Chapter Review, Page 171

1. Linear   3. Ordered pairs   5. Are on the same line   7. Slope   9. No slope
11. Lines cross at one point   13. No

CHAPTER 13

Page 178

1. (a) $(1, 1), (1, 2), (1, 3), (1, 4), (2, 1), (2, 2), (2, 3)$;
   (b) $(1, 5), (2, 4), (2, 5), (3, 2), (3, 3), (3, 4), (3, 5), (4, 1), \ldots, (4, 5), (5, 1), \ldots,$
   $(5, 5)$;   (c) Same as (a) including $(2, 3)$;   (d) Same as (b) including $(2, 3)$;
   (e) All points;   (f) No points;   (g) All points;   (h) No points;   (i) No points;
   (j) All points
3. H   5. A   7. F   9. C
11. (a)

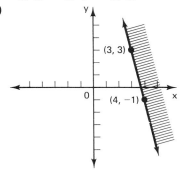

(b)

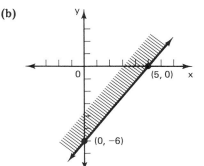

(c)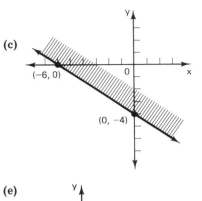

(d)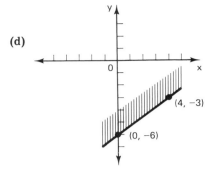

(e)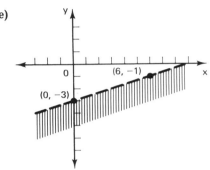

(f)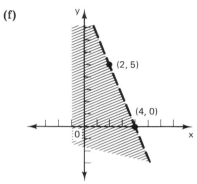

Page 188

1. C   3. A

5. **(a)**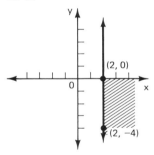

(b)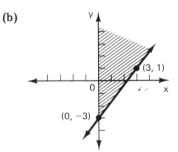

(c)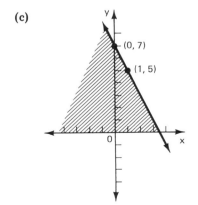

(d)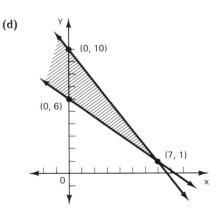

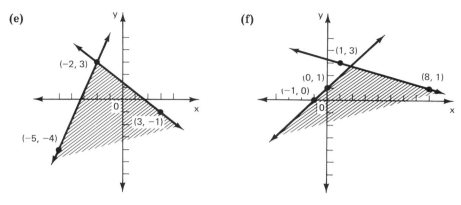

(e)   (f)

7. (a) $(0, 7), (7, 0), (11, 3)$;   (b) $(10, 0)$ $(9, 8), (14, 0)$;
   (c) $(2, 3)$ $(0, 0), (10, 0), (13, 6)$;   (d) $(11, 0), (0, 0), (0, \frac{4}{7}), (\frac{377}{47}, \frac{350}{47})$

Chapter Review, Page 191

1. Conditional   3. Half; $3x + 4y = 12$ 5. Polygon   7. Maximum; minimum
9. Linear programming

CHAPTER 14

Page 195

1. (b); (c); (d)   3. $\frac{2}{7}, \frac{4}{7}, 1, \frac{9}{7}, \frac{10}{7}$
5. (a) 10;   (b) 17;   (c) $8 - 2a$;   (d) $5b - 3$;   (e) -15;   (f) $5c - 2d + 15$;   (g) 4.5

Page 199

1. (a) -2;   (b) 16;   (c) $x^2 + x + 8$;   (d) $6 + 5x - x^2$;   (e) 18;   (f) 5;
   (g) $x^2 - 5x - 6$;   (h) $12 + 5a - a^2$
3. (a) $21x^2 + 10x - 16$;   (b) $5x + 3$;   (c) -49;   (d) $\frac{22}{10}$;   (e) $12x^2 + 22x - 20$;
   (f) $\frac{10}{22}$
5. May vary

Chapter Review, Page 200

1. Relation   3. Not; 2 has two values: 4 and 7   5. Range   7. 2; 3
9. Composite

CHAPTER 15

Page 206

1. (a) $5, 0, -5$;   (b) $21, 25, 29$;   (c) $-32, 64, -128$;   (d) $63, 127, 255$;
   (e) $-21, -33, -47$

3. **(a)** 3, 7, 11, 15, 19;   **(b)** 1, 2, 4, 8, 16;   **(c)** 8, 12, 8, 12, 8;
   (d) 2, 4, 16, 256, 65536;   **(e)** $-5, 120, -380, 1620, -6380$;
   (f) 100, 50, 25, 12.5, 6.25;   **(g)** 100, 20, 4, .8, .16;   **(h)** 1, 1, 2, 3, 5;
   (i) 1, 2, 2, 4, 8

5. **(a)** $S(1) = 5$
       $S(N) = S(N - 1) + 5$
   (b) $S(1) = 2$
       $S(N) = S(N - 1) + 3$
   (c) $S(1) = 1$
       $S(N) = 2 * S(N - 1)$
   (d) $S(1) = 2$
       $S(N) = 3 * S(N - 1)$

   (e) $S(1) = 3$
       $S(N) = S(N - 1) + 5 * 2 * * (N - 2)$
   (f) $S(1) = 4$
       $S(N) = S(N - 1) + 2 * * (N - 2)$
   (g) $S(1) = 1$
       $S(N) = -1 * S(N - 1)$
   (h) $S(1) = 3$
       $S(N) = S(N - 1) * * 2$

7. 7; 16; 19
   $S(1) = 1$
   $S(N) = S(N - 1) + 3$
   $SN = 3N - 2$

Page 211

1. **(a)** 8, 10, 12, 14, 16;   **(b)** 5, 8, 11, 14, 17;   **(c)** $-3, 1, 5, 9, 13$;
   (d) 225, 230, 235, 240, 245;   **(e)** $-\frac{1}{2}, \frac{3}{2}, \frac{7}{2}, \frac{11}{2}, \frac{15}{2}$;
   (f) $\frac{1}{2}, \frac{3}{4}, \frac{4}{4}, \frac{5}{4}, \frac{6}{4}$;   **(g)** $\frac{1}{3}, \frac{3}{3}, \frac{5}{3}, \frac{7}{3}, \frac{9}{3}$;   **(h)** 100, 108, 116, 124, 132;
   (i) 2, $-2, -6, -10, -14$;   **(j)** 24, 18, 12, 6, 0

3. $S(1) = 1$
   $S(N) = S(N - 1) + 3.5$

   $S(1) = 24$
   $S(N) = S(N - 1) - 3$

   $S(1) = -8$
   $S(N) = S(N - 1) + 2.48$

5. Step 1:   Set N = 1
   Step 2: $I = -10$
   Step 3: $S(N) = I$
   Step 4: $N = N + 1$
   Step 5:   Is N $>$ 50?
               Yes: Go to step 6
               No: $I = I + .25$
                   Go to step 3
   Step 6:   Stop

7. Step 1:   Set N = 1
   Step 2: $I = 100$
   Step 3: $S(N) = I$
   Step 4: $N = N + 1$
   Step 5: $I = I - 2$
   Step 6:   Is I $<$ -96?
               Yes: Go to step 7
               No: Go to step 3
   Step 7:   Stop

9. 492

Page 215

1. (a); (b); (d); (e)
3. **(a)** 3;   **(b)** .0013122
   (c) $S(1) = 6$
       $S(N) = S(N - 1) * 3$
   (d) Step 1: $N = 1$
       Step 2: $1 = 6$
       Step 3: $S(N) = I$
       Step 4: $I = I * .3$
       Step 5: $N = N + 1$
       Step 6: Is $N > 100$?
                 Yes: Go to step 7
                 No: Go to step 3
       Step 7: Stop
   (e)

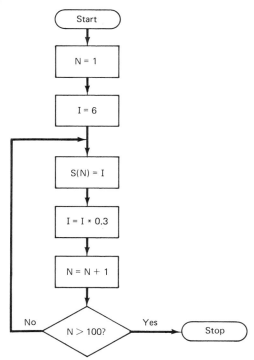

   (f) $S(10)$
5. 13

Page 218

1. **(a)** 1, 119;   **(b)** 5, 104;   **(c)** 5, 675

3.

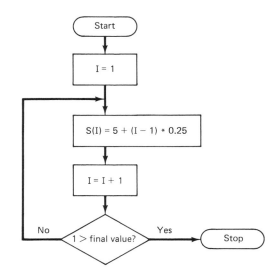

5.

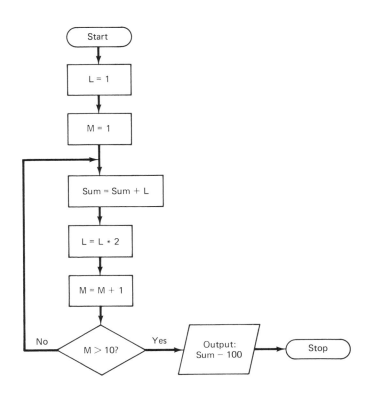

Chapter Review, Page 221

1. Integers   3. $1; 2N - 1$ 5. Common difference
7. $S(1) = A, S(N) = S(N - 1) + D$ 9. $A(R)^{N-1}$

CHAPTER 16

Page 228

1. **(a)** $(3, 4)$; **(b)** $(2, 2)$; **(c)** $(4, 2)$; **(d)** $(3, 2)$; **(e)** $(2, 6)$; **(f)** $(5, 3)$
3. **(a)** 25; **(b)** 43; **(c)** 81; **(d)** 72; **(e)** 59; **(f)** 53; **(g)** 93; **(h)** 68;
   (i) 62; **(j)** 69
5.

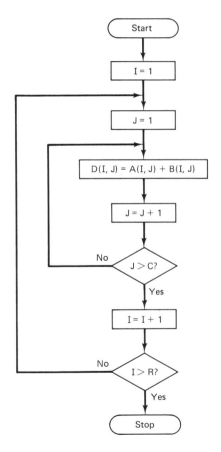

7. **(a)** $(3, 2)$; **(b)** $(2, 5)$; **(c)** $(7, 6)$; **(d)** $(9, 8)$; **(e)** $(2, 2)$; **(f)** $(6, 10)$;
   (g) $(5, 11)$; **(h)** $(6, 2)$; **(i)** $(7, 2)$; **(j)** $(5, 5)$

Page 233

3. **(a)** $\begin{bmatrix} -\dfrac{2}{13} & \dfrac{3}{13} \\[2mm] \dfrac{7}{13} & -\dfrac{4}{13} \end{bmatrix}$ **(b)** $\begin{bmatrix} \dfrac{20}{109} & -\dfrac{23}{109} \\[2mm] -\dfrac{17}{109} & \dfrac{25}{109} \end{bmatrix}$ **(c)** $\begin{bmatrix} \dfrac{9}{432} & \dfrac{6}{432} \\[2mm] \dfrac{36}{432} & -\dfrac{24}{432} \end{bmatrix}$ **(d)** $\begin{bmatrix} \dfrac{42}{1776} & -\dfrac{48}{1776} \\[2mm] -\dfrac{75}{1776} & \dfrac{128}{1776} \end{bmatrix}$

   (e) No inverse

5. **(a)** $(2, 4)$; **(b)** $(5, 2)$; **(c)** $(3, 6)$; **(d)** $(3, 3)$; **(e)** $(4, 4)$

Page 241

1. **(a)** $(3, 4)$;

   (b) $\begin{bmatrix} 2 & 3 & 4 \\ 3 & 5 & 7 \\ 4 & 7 & 10 \\ 4 & 8 & 12 \end{bmatrix}$

3.

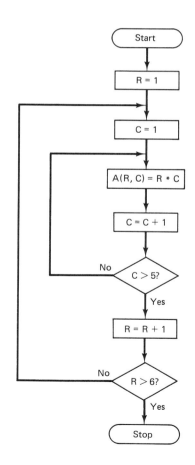

5.

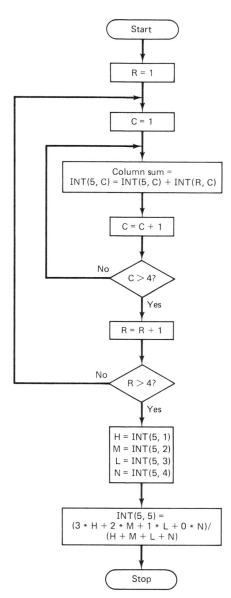

Chapter Review, Page 244

1. 3   **3.** 5; 3   **5.** 1; 2; 3; 4; 5; 6

7. (a) $\begin{bmatrix} 11 & 11 \\ 11 & 13 \end{bmatrix}$ **(b)** $\begin{bmatrix} 7 & -3 \\ -7 & -11 \end{bmatrix}$ **(c)** $\begin{bmatrix} 27 & 12 \\ 6 & 3 \end{bmatrix}$ **(d)** $\begin{bmatrix} -4 & -14 \\ -18 & -24 \end{bmatrix}$

9. M; R   **11.** 1; 0   **13.** The identity matrix   **15.** Two dimensional

CHAPTER 17

Page 254

1. (a) $5 \times 10^2 + 7 \times 10^1 + 8 \times 10^0$;
   (b) $1 \times 4^3 + 3 \times 4^2 + 2 \times 4^1 + 1 \times 4^0 + 2 \times 4^{-1} + 2 \times 4^{-2} + 1 \times 4^{-3}$;
   (c) $4 \times 5^1 + 3 \times 5^0 + 4 \times 5^{-1} + 2 \times 5^{-2} + 2 \times 5^{-3}$;
   (d) $1 \times 9^2 + 7 \times 9^1 + 8 \times 9^0 + 7 \times 9^{-1} + 7 \times 9^{-2}$;
   (e) $2 \times 3^3 + 2 \times 3^2 + 1 \times 3^1 + 2 \times 3^0 + 2 \times 3^{-1} + 2 \times 3^{-2} + 3 \times 3^{-3}$

3. (a) 3, (22) base 3; (b) 9, (121) base 9; (c) 6, (14) base 6; (d) 7, (45) base 7;
   (e) 5, (132) base 5

5.

100000	32768	16807	7776	243
10000	4096	2401	1296	81
1000	512	343	216	27
100	64	49	36	9
10	8	7	6	3
1	1	1	1	1
.1	.125	.14286	$.1\overline{6}$	$.\overline{3}$
.01	.015625	.020408	$.02\overline{7}$	$.\overline{1}$
.001	.00195313	.00292	.0046296	$.\overline{037}$
.0001	.00024414	.000416	.0007716	.01234568

7. (a) (112000) base 3; (b) (1430) base 6; (c) (3003) base 5; (d) (460) base 9;
   (e) (101111010) base 2

Page 259

1. (a) $(.30\overline{6314})$ base 8; (b) $(.1\overline{4320})$ base 5; (c) $(.\overline{1011})$ base 3;
   (d) $(.\overline{2466})$ base 7; (e) $(.\overline{34})$ base 9

3. (a) $(3434.2421131\ldots)$ base 5; (b) (756.45) base 8;
   (c) $608.5174088\ldots)$ base 9; (d) (111101110.100101) base 2;
   (e) $(1304.4022042\ldots)$ base 7

5. $(2021221112.102111111220)$ base 3

7. $(1110010010000001.100110011111)$ base 2

Page 267

1. (a) (113) base 5; (b) (1314) base 5; (c) (33004) base 5; (d) (134) base 5;
   (e) (21232) base 5

3. (a) (13137) base 9; (b) (175143) base 9

5. (a) $100, 101, 102, 103, 110$; (b) $122, 200, 201, 202, 210$;
   (c) $65, 66, 100, 101, 102$; (d) $140, 141, 142, 143, 144$

7. (a) 414667; (b) 012554; (c) 013341; (d) 1032211; (e) 440344544

Page 274

1.

0	0	0	0
0	1	2	3
0	2	3	12
0	3	12	21

3. (a) (120102) base 3;   (b) (2121210) base 4;   (c) (6837183) base 9;
   (d) (511544) base 6
5. (a) (2122) base 3;   (b) (43142) base 5;   (c) (211463) base 9;
   (d) (52405.$\overline{3}$) base 6;   (e) (326022 R 45) base 7

Chapter Review, Page 276

1. 10^2 3. 4; (23) base 4
5. (a) Expanded form;   (b) Multiply, place value;   (c) Add
7. {0, 1, 2, 3}   9. 9; remainders   11. 4; .12; 3   13. 2010; 3's
15. Addition   17. Multiplication

CHAPTER 18

Page 284

1. (a) 107;   (b) 7281;   (c) 170.125;   (d) 13.78125;   (e) 7.875;   (f) 467;
   (g) 84.125;   (h) 5.8007813;   (i) 56.797119;   (j) 0.1630859
3. (a) (100010) base 2;   (b) (111) base 2;   (c) (100101) base 2;   (d) –(1101) base 2
5. (a) (11011) binary, (33) octal;   (b) (10000100) binary, (204) octal;
   (c) (11111111) binary, (377) octal;   (d) (.01$\overline{1100}$) binary, (.3$\overline{4631}$) octal;
   (e) (101110100.100001111010 . . .) binary, (564.41727) octal
7. (a) (100111101110010) binary;   (b) (1011000100.101) binary;
   (c) (101111.100011011) binary;   (d) (1010010.001110) binary;
   (e) (.111101100010) binary

Page 289

1. A   64
   14   C 8
   1 E   12 C
   28   190
   32   F 14
   3 C   258
   46   2 BC
   50   320
   5 A   384
3. (a) 142;   (b) 8204;   (c) 266.6875;   (d) 67.072754;   (e) 42355.688
5. (a) (15.73) octal;   (b) (100110111.101101) binary;
   (c) (1101000110100.1011) binary;   (d) (167.481) hexadecimal;
   (e) (EA.58) hexadecimal
7. (a) (10343) hexadecimal;   (b) (98) hexadecimal;   (c) (114350) hexadecimal
9. (a) (785C) 15's complement;   (b) (654B) 15's complement;
   (c) (EOB87) 15's complement;   (d) (AO721) 15's complement;
   (e) (ECBDA) 15's complement

Chapter Review, Page 292

1. Binary   3. 16   5. Positive   7. Three   9. 16

CHAPTER 19

Page 298

1. (a) 001110010101;   (b) 001110100101;   (c) 001111110101;
   (d) 011011001000;   (e) 110001011
3. (a) 0111 0111;   (b) 0111 0111;   (c) 1101 1101;   (d) 1010   1010;
   (e) 1001101
5. 0   0000
   1   0001
   2   0010
   3   0100
   4   0101
   5   1010
   6   1011
   7   1101
   8   1110
   9   1111
7. $(4)(4)(3)(-2)$; $(8)(4)(-2)(-1)$
9. (a) 0001 1101;   (b) 0001 1010;   (c) 0011 1000;   (d) 0011 1010;
   (e) 0111 1001

Page 301

1. F0   F8   C7   D6   F6   5B
   F1   F9   C8   D7   F7   5C
   F2   C1   C9   D8   F8   5D
   F3   C2   D1   D9   F9   6O
   F4   C3   D2   F2   4O   6I
   F5   C4   D3   F3   4B   6B
   F6   C5   D4   F4   4D   7E
   F7   C6   D5   F5   4E
3. (a) 00000000011111000100000000010011000100011100 0110;
   (b) 0007610004610706;
   (c) 011000001101111010111101100001101000101111011 01110110110;
   (d) 0101000001011011010011110101000001010100010011 1101010-
       11101010110;   (e) 50574750544F5756;
   (f) 1111000011101110110000111110000111101000110000 111110-
       11111110110;   (g) F0F761F0F461F7F6

Chapter Review, Page 303

1. 4   3. 15   5. Self-complementing   7. Fewest 1's   9. 7

CHAPTER 20

Page 310

1. (a) $2^4 = 16$
   (b) TTTT   FTTT
        TTTF   FTTF
        TTFT   FTFT
        TTFF   FTFF
        TFTT   FFTT
        TFTF   FFTF
        TFFT   FFFT
        TFFF   FFFF

3. | (a) | (b) | (c) | (d) | (e) | (f) | (g) | (h) |
   |-----|-----|-----|-----|-----|-----|-----|-----|
   | T | T | T | T | T | T | T | T |
   | T | F | T | T | T | F | F | T |
   | F | T | T | T | T | T | T | T |
   | F | F | T | T | F | F | F | F |
   | F | F | T | T | T | T | T | T |
   | F | F | T | F | F | F | F | F |
   | F | F | F | T | T | F | F | T |
   | F | F | F | F | F | F | F | F |

5. | | | | |
   |---|---|---|---|
   | T | T | F | F |
   | T | F | F | T |
   | F | T | T | F |
   | F | F | T | T |

7. OR of two negatives: $\sim(P \wedge Q) = \sim P \vee \sim Q$
   AND of two negatives: $\sim(P \vee Q) = \sim P \wedge \sim Q$

Page 316

1. (a) *If* you worked for one year, *then* you are eligible for a raise;
   (b) *If* $3x + 5 = 14$, *then* $x = 3$;
   (c) *If* there is a power loss, *then* the lights will go off;
   (d) *If* a polygon has three sides, *then* it is a triangle;
   (e) *If* you apply for a permit, *then* you must be 16 or older;
   (f) *If* you have no license, *then* you cannot drive a car.

3. (a) T; (b) F; (c) T; (d) T; (e) F

5. | (a) | (b) | (c) | (d) | (e) | (f) | (g) |
   |-----|-----|-----|-----|-----|-----|-----|
   | T | T | T | T | T | T | F |
   | T | F | T | F | F | T | T |
   | T | T | T | T | T | F | T |
   | T | T | T | T | T | T | F |

Page 319

1. D   3. E   5. A   7. A   9. A   11. A   12. C   15. B

Chapter Review, Page 321

1. Negative
3. **(a)** And; **(b)** Or; **(c)** Not; **(d)** Conditional: if, then;
   (e) Biconditional: If and only if
5. Either
7. **(a)** $P \rightarrow \sim Q$; **(b)** $Q \rightarrow \sim P$; **(c)** $\sim Q \rightarrow P$
9. Biconditional   11. Tautology

CHAPTER 21

Page 326

1. **(a)** 1 **(b)** 1
       1   1
       1   1
       0   1
       0   1
       0   0
       0   0
       0   0
3. **(a)** T; **(b)** F; **(c)** F; **(d)** T; **(e)** T; **(f)** T; **(g)** T
5. **(a)** $(A \bullet B) + (\overline{A} \bullet C)$; **(b)** $(X + Y) \bullet (W + Z) \bullet (\overline{Z} + \overline{Y})$

Page 333

1. $F(X, Y) = XY + X\overline{Y}$
3. **(a)** $(\overline{X} + \overline{Z} + \overline{Y}) (\overline{X} + Y + Z) (X + \overline{Y} + \overline{Z}) (X + Y + \overline{Z})$;
   (b) $XY\overline{Z} + X\overline{Y}Z + \overline{X}Y\overline{Z} + \overline{X}\overline{Y}\overline{Z}$
5. **(a)** $XYZ + X\overline{Y}Z + \overline{X}Y\overline{Z} + \overline{X}\overline{Y}Z$;
   (b) $(\overline{X} + \overline{Y} + Z) (\overline{X} + Y + \overline{Z}) (X + \overline{Y} + Z) (X + \overline{Y} + Z)$;
   (c) $(\overline{X} + \overline{Y} + \overline{Z}) (\overline{X} + Y + Z) (X + Y + \overline{Z}) (X + Y + Z)$;
   (d) $(XY\overline{Z}) + (X\overline{Y}Z) + (\overline{X}YZ) + (\overline{X}Y\overline{Z})$
   (e)

(f)

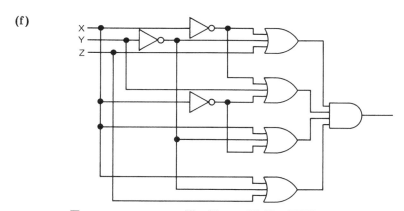

7. **(a)** $XY + \overline{X}Y$;   **(b)** $XYZ + XY\overline{Z} + \overline{X}YZ + \overline{X}Y\overline{Z} + \overline{X}\overline{Y}Z$;
      (c) $X\overline{Y}Z + X\overline{Y}\overline{Z} + \overline{X}YZ + \overline{X}Y\overline{Z} + \overline{X}\overline{Y}Z$;   **(d)** $XYZ + XY\overline{Z} + X\overline{Y}Z + \overline{X}YZ$;
      (e) $XYZ + XY\overline{Z} + X\overline{Y}Z + X\overline{Y}\overline{Z} + \overline{X}YZ + \overline{X}\overline{Y}Z$

Page 344

1.

$\overline{XYWZ}$	$\overline{X}\overline{Y}\overline{W}Z$	$\overline{X}Y\overline{W}Z$	$\overline{X}YW\overline{Z}$
$X\overline{Y}W\overline{Z}$	$X\overline{Y}\overline{W}Z$	$X\overline{Y}\overline{W}Z$	$XY\overline{W}Z$
$X\overline{Y}W\overline{Z}$	$X\overline{Y}WZ$	$XYWZ$	$XYW\overline{Z}$
$\overline{X}YW\overline{Z}$	$\overline{X}Y\overline{W}Z$	$\overline{X}YWZ$	$\overline{X}YW\overline{Z}$

3. **(a)** $\overline{X} + \overline{Y} + \overline{Z}$;   **(b)** X;   **(c)** $Z + XY$;   **(d)** $YZ + X\overline{Z}$;   **(e)** $\overline{W}Z + \overline{W}Y + \overline{W}X$
5. **(a)** $\overline{X}Z$;   **(b)** $\overline{X} + \overline{Z}$;
     (c) 0
        1
        0
        1
        1
        1
        1
        1

Page 348

1. **(a)**

1	1	1	1	1
1	1	0	1	0
1	0	1	1	0
1	0	0	0	1
0	1	1	1	0
0	1	0	0	1
0	0	1	0	1
0	0	0	0	0

     (b) Carry = $XC + XY + YC$
          Sum $:= XYC + X\overline{Y}\overline{C} + \overline{X}Y\overline{C} + \overline{X}\overline{Y}C$

(c)

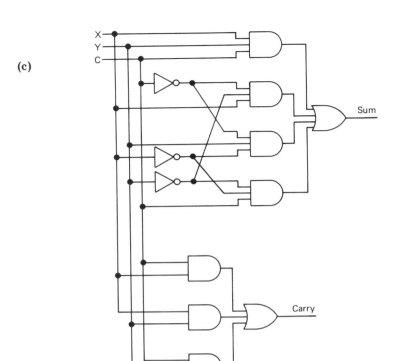

(d) $(\overline{X} + \overline{Y} + C)(\overline{X} + Y + \overline{C})(X + Y + C)(X + \overline{Y} + \overline{C}) = \text{sum}$
$(X + Y)(Y + C)(X + C) = \text{carry}$

(e)

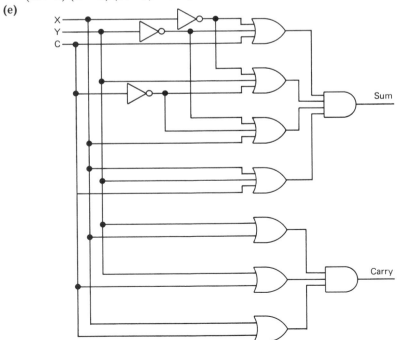

384

3. (a) $\overline{X}B + \overline{X}Y + YB$ = borrow
$XYB + X\overline{Y}\,\overline{B} + \overline{X}\,Y\overline{B} + \overline{X}\,\overline{Y}B$ = difference

(b)

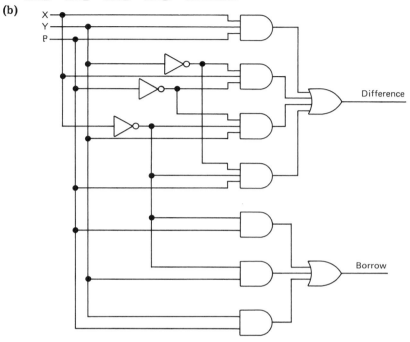

Chapter Review, Page 349

1. Product; Sum; Negation
3. 0; 1
5. $X \cdot Y + X \cdot Z$; $(X + Y) \cdot (X + Z)$
7. E
9. D
11. $\overline{W}Z$

Index